# Beyond the Checkride

## Other McGraw-Hill Books of Interest

*Flight Test Tips and Tales from the Eye of the Examiner.*
Howard Fried. 0-07-022462-5

*Piloting for Maximum Performance.*
Lewis Bjork. 0-07-005699-4

*Spin Management & Recovery.*
Michael C. Love. 0-07-038810-5

*Handling In-Flight Emergencies.*
Jerry A. Eichenberger. 0-07-015093-1

*Cockpit Resource Management.*
Thomas P. Turner. 0-07-065604-5

*Stick and Rudder.*
Wolfgang Langewiesche. 0-07-036240-8

*Avoiding Mid-Air Collisions.*
Shari Stamford Krause. 0-07-035945-8

*Understanding Aeronautical Charts.*
Terry T. Lankford. 0-07-036467-2

*Private Pilot Test Guide & Disk.*
Douglas S. Carmody. 0-07-912308-2

*Instrument Pilot Test Guide & Disk.*
Douglas S. Carmody. 0-07-011611-X

*Commercial Pilot Test Guide.*
Douglas S. Carmody. 0-07-011519-2

*Companion disk sold separately via coupon included in book.*

# Beyond the Checkride

*Howard Fried*

**McGraw-Hill**

New York  San Francisco  Washington, D.C.  Auckland  Bogotá
Caracas  Lisbon  London  Madrid  Mexico City  Milan
Montreal  New Delhi  San Juan  Singapore
Sydney  Tokyo  Toronto

**Library of Congress Cataloging-in-Publication Data**

Fried, Howard
  Beyond the checkride : what your instructor never taught
you / by Howard Fried
      p.    cm.
  Includes index.
  ISBN 0-07-022468-4 (pbk.)
  1. Airplanes—Piloting.  I. Title.
TL710.F7597  1997
629.132'5217—dc21                                    97-14114
                                                        CIP

# McGraw-Hill

*A Division of The McGraw·Hill Companies*

7 8 9 10   QPF/QPF   0 5 4 3 2

ISBN 0-07-022468-4

*The sponsoring editor for this book was Shelley Chevalier, the editing super-
visor was Sally Glover, and the production supervisor was Tina Cameron. It
was set in Garamond by McGraw-Hill's desktop publishing department in
Hightstown, N.J.*

*Printed and bound by Quebecor/Fairfield.*

This book is printed on recycled, acid-free paper containing
a minimum of 50% recycled, de-inked fiber.

McGraw-Hill books are available at special quantity discounts to use as pre-
miums and sales promotions, or for use in corporate training programs. For
more information, please write to the Director of Special Sales, McGraw-Hill,
11 West 19th Street, New York, NY 10011. Or contact your local bookstore.

# Contents

# Contents

# Preface

I've been writing the "Eye of the Examiner" column for *FLYING* magazine for over six years now, and my correspondents, plus my observations as a flight instructor and pilot examiner over more years than I care to contemplate, have convinced me that there is a veritable mountain of material that pilots must learn after their flight instructors have passed them on into the world of the certificated pilot. This knowledge can only come from experience, and as we all know, experience can't be rushed. It only comes with the passage of time, lots of time. There is, however, a means by which we can take a shortcut to gaining experience. That is by taking advantage of the experience of others. This is the easy way, and being lazy, I like to do things the easy way. How about you?

If you desire to benefit from the experience of hundreds of others who have gone before, this book is for you. If you feel the need to experience it all for yourself, stop wasting your time reading this right now and go out and start spending the next few hundred years acquiring the experience offered here. I make no claim to be an expert aviator, but I have accumulated in excess of 40,000 hours in the air in a career that has spanned more than half a century, and I'm still here, so I must be doing something right. As a pilot I'm just an average airplane manipulator, but I do try to take advantage of every learning experience to which I'm exposed. How about you? Do you seize every opportunity to learn from mistakes, both yours and others? If not, you are wasting a valuable chance to learn from experience. Here, then, are a whole bunch of interesting, educational, and entertaining tidbits I have acquired over the years.

# Acknowledgments

Much of that portion of Chapter 1 dealing with prop safety is taken from an article I wrote which was published in the March, 1996 issue of *USAviator* magazine and in the *Sport Plane Resource Guide (Second Edition)*. Our thanks to publisher Jim Campbell for granting us permission to reprint it here.

The drawings in Chapters 3 and 4 were taken from slides I use in a presentation dealing with some of the hazardous attitudes that get pilots in trouble. These slides were made for me by ICOM, Inc. of Columbus, Ohio, and Southfield, Michigan, and graciously presented to me by Phil Yoder, its president and CEO. Mr. Yoder manages both his Columbus and Southfield offices by means of traveling between them in his Cessna 310.

Much of the material in the section of Chapter 3 on crash survivability was supplied by St. Elmo, "Buz" Massengale, Safety Program Manager at the Jackson, Mississippi, DOT-FAA Flight Standards District Office (FSDO). He is also responsible for making some of the information which is included in Chapter 11 available to me, as well as that portion of Chapter 5 dealing with stress. Thanks, Buz.

Unlike many of the people in the Flight Standards Division of the FAA who are responsible for the old joke that goes, "Hi, I'm from the FAA and I'm here to help you," when the guy is really there to hang you, Buz is actually there to help us, the users of the system and the occupiers of the airspace. The real public servants in the FAA consider us as customers and treat us as such, rather than adversaries to be defeated given the slightest opportunity, as so many in the Flight Standards Division of the FAA are prone to do. Buz, only one "Z" (he must have lost one someplace), is a member of that rare breed, a true public servant in the highest sense, giving of himself unstintingly in the service of the flying public. He makes himself and his resources available at any time to anybody. His information is always current and accurate.

Part of the section in Chapter 13 dealing with the Remedial Training Program of the FAA was taken from an article I wrote at the behest of Buz Massingale, which was published in *USAviator* magazine, reprinted in the *Journal of the Lawyer Pilots Bar Association*, and again in my book, *Flight Test Tips and Tales from the Eye of the Examiner*, published in 1996 by McGraw-Hill.

Thanks are also due Ms. Michelle Panabecker-Neff, whose original post on America Online quoted in Chapter 15 started a raging controversy over the validity of ab initio flight training, which is currently being tested by some air carriers and by the military. Due to a problem with her computer program, Ms. Penabecker-Neff had a problem expressing herself, and she has been consistently unfairly criticized and attacked for her spelling and grammar, rather than for her ideas. This is indeed regrettable since she opened a worthwhile discussion in an area in which there is room for legitimate disagreement.

I am also indebted to Dick Knapinski of the EAA. Dick does public relations work for the Experimental Aircraft Association in Oshkosh, Wisconsin, and he graciously supplied some of the material for that part of Chapter 16 that deals with the media, how aviation is treated by the popular media, and what can be done about it. Mr. Knapinski made his extensive file on the subject available, and I drew heavily on it in my presentation of the subject. Thanks, Dick!

Myrna Papurt, DVM, was very helpful in supplying me with some of the information in Chapter 7 on the subject of traveling with pets. Thanks, Myrna.

All of us who enjoy a good story owe special thanks to Jerry Temple, who provided us with the helicopter story in Chapter 10. I'm sure you'll get a kick out of it, as I have.

Finally, I would certainly be remiss if I failed to acknowledge the fact that Russ (Razz) Glover, a superb pilot and excellent flight instructor, drew on his vast experience to read and critique several chapters of this work. Razz is a very bright guy who has been instructing for a great many years. He has a wealth of knowledge in the field of aviation education, and his suggestions have been invaluable in the preparation of this work. Of course, any mistakes you may find here are mine alone.

No matter how carefully one researches an effort such as this, in so controversial an area there are bound to be disagreements, and if you should disagree with anything you find here, please bear in mind that there is plenty of room for honest differences of opinion. The entire area of aviation education is a controversial one.

I must also thank Brian Jacobson for the generous use of his fancy computer equipment in the preparation of the manuscript, as well as his counsel on some of the material contained herein.

Like any writer, I am especially grateful to my wife for her patience and understanding while I spent countless hours at the word processor, ignoring the household chores, which remained undone.

# Introduction

The standard for any certificate or rating as set forth by the FAA in the various Practical Test Standards is the absolute minimum of performance and knowledge to which an applicant may be held. Thus any given certificate or rating is truly, as they have often been called, "a license to learn." This is true in spite of the fact that a pilot certificate in the strictest sense isn't a license at all. I have been flying for well over 50 years, and every time I get in an airplane and take off I learn something new, and if this ever quits being the case, then I truly believe it will be time to quit doing it.

*FLYING* magazine has for many years published a regular monthly feature called "I Learned about Flying from That," and although a few are submitted by student pilots, the majority of these columns deal with principles learned by certificated pilots, sometimes long after they have earned their certificates and ratings.

Over the years I have received numerous suggestions to write a book about valuable information that is frequently left out or brushed over during flight training. However, what really inspired me to sit down and begin to put some of these things on paper, what finally triggered the present effort, was an incident which occurred not long ago while I was sitting in the right seat of a light twin being flown by an individual who holds a commercial pilot certificate with multiengine and instrument privileges. We were VFR at 8,500 feet northwestbound on a more than 300-nautical-mile leg. The wind had been forecast right on our nose at 50 knots at six and 55 knots at niner thousand. My companion turned to me and said, "Let's see how accurate the winds aloft forecast is."

"Okay," I responded.

He then cranked the Loran C around to read the ground speed, and he compared that reading to what the airspeed indicator was showing and remarked, "They're off by about 10 knots."

"Wait a minute," I said. "You're reading indicated airspeed. This airplane has a true airspeed indicator. To make a valid comparison, you have to first get the true airspeed."

At this point he informed me that he didn't know how to work the true airspeed indicator. He pointed out that nobody had shown him how to use the instrument and he had read nothing about it. I then showed him that if he reset the altimeter to obtain pressure altitude, then rotated the little knob to line up the pressure altitude against the outside air temperature, he could read the true airspeed right off the little window at the bottom of the airspeed indicator. Of course, had he been trained in the use of a good old E6B (or even a D4 type computer) instead of one of those newfangled electronic flight computers, he would have recognized the true airspeed indicator and known how it works without further instruction. Since this is a really bright guy, I was surprised that he hadn't figured it out for himself.

Since that time I have asked several fairly high-time pilots to explain the procedure for obtaining true airspeed from a true airspeed indicator, and not a single one was able to do so. I guess it is so simple a procedure that most flight instructors, believing it to be self-explanatory, don't bother to explain it to their students. Virtually all of those pilots who did know the procedure had figured it out for themselves.

It is not particularly difficult to understand how a person, even a person with a commercial multiengine instrument certificate, and a particularly bright fellow at that, could have missed learning this simple technique. It is something most flight instructors expect their students to pick up on their own. After all, the technique for using a true airspeed indicator is more or less self-explanatory. When I checked with a few instructors, I found none that made a point of teaching their students the use of this tool. I also learned that many of today's advanced students don't know how to get the true airspeed from a "whiz wheel" (E6B computer), and if the battery failed on their hand-held electronic flight computer, they'd be hopelessly lost. These same pilots, of course, also lack the knowledge to properly use and read a true airspeed indicator, as well as several other things that professional pilots take for granted. In administering some 4,000 flight tests and in interviewing several dozens of instructors who specialize in advanced training, I learned a lot about the amount of material that pilots, good pilots, failed to learn during the training process.

And there are several specific areas in which a large number of certificated private pilots demonstrate a lack of ability in performing the most fundamental of tasks. More than one instructor has told me of private pilots who were incapable of tuning and following a VOR. The most extreme case of this is in a story related by Rod Machado, who told me of a pilot who, when asked on a flight test to tune and fly to a VOR, tuned the station, centered the needle in the OBS, and then rotated the heading indicator until the same number as shown on the OBS (omni bearing selector) showed up. When asked what he was doing, he replied, "My instructor told me that the DG should always match the VOR indication." This tale may seem amusing, but it is also tragic in its implications.

Added to these performance areas in which a large number of certificated pilots find holes in their knowledge is another area in which a great many pilots are lacking knowledge through no fault of their own, or even of their instructors. Recently a substantial amount of emphasis has been placed on human factors in the prevention of aircraft accidents. This, then, indicates another area that really needs to be covered in a book such as this. Obviously, if a majority of aviation mishaps are caused by human error, this is an area that definitely requires attention.

The vast majority of general-aviation pilots, even those with hundreds of hours of experience, have never been anywhere near the outer limits of the performance envelope of the airplanes they fly. Why not? Because their instructors never showed them all that the training airplane can do, and believe me, it is a whole lot! This may be because many instructors have themselves never explored the limits of the performance of the training airplanes in which they teach, and they can't very well demonstrate that which they've never done themselves.

Consequently, I finally decided the time has come for me to reduce to writing some of the principles that I have learned in my more than 50 years and 40,000 hours of flying. I am by no means alone in this effort. I have consulted several friends who have been flying for a great many years, involving many thousands of hours of experience, and I am herewith offering some of the wisdom that has been garnered over several hundreds of years and megathousands of hours of experience accrued by myself and others.

None of this material is new. Much of it will be familiar to many of those who read it. But if every reader manages to glean a single fact or principle that he or she didn't know or has forgotten, and that proves helpful, it will be well worth the purchase price of this little book.

It is just not possible for a flight instructor, any flight instructor, to teach his or her students how to react to every problem they might encounter as pilots. However, if we can teach them to understand, through analysis, the problem, then they can work out their own, correct solutions.

Starting with the premise that virtually no pilots sit down in an airplane with the intention of killing themselves, here then are some principles that may help to prevent that undesirable result.

Finally, a word about style. If you think there are too many "his/her," and "he or she" expressions in this work, there's a reason. Since the English language has no word for a gender-neutral situation, proper usage has always been to use the word "he" or "him" when referring to both sexes. Now, however, that this is no longer acceptable, so I use "he or she" and "him/her." Note: I do not refer to a person as a chair. A chair is something you sit on. If it is a female, she can be a chairwoman or chairperson, but definitely not a chair!

# 1

# Basics rarely taught completely

## I Maneuvering speed

It's not marked on the airspeed indicator, although it may be placarded on the panel, and you may have to search through the
*Approved Flight Manual* to find it, but one of the very most important numbers for the pilot to know is Va, or maneuvering speed. It
behooves every pilot to know and thoroughly understand maneuvering speed and why it is so important to go right to this speed
when turbulence is encountered.

I know a flight instructor who has, for several years, taught his students to slow down in rough air by explaining that it works the same
way as an automobile going down a rough road or crossing a set of
bumpy railroad tracks. If the car is slowed down, it will ride the
bumps, and although the ride may not be smooth, no damage will
result. Conversely, if the driver takes the bumps at a high rate of
speed, something is likely to break, altogether a clever way of expressing what may happen. While this unique instruction technique
may accomplish the purpose, it fails to properly explain what maneuvering speed is and why it works to prevent structural damage.

Sooner or later every pilot becomes acquainted with the definition of
maneuvering speed as the maximum speed at which the pilot may
make an abrupt, full-control deflection without causing structural
damage. When reference is made to control deflection it means all
the controls, throttle, flaps, ailerons, rudder, and elevator. Which one
is being emphasized? The elevator, of course. Students are told to go
to and maintain Va whenever they encounter turbulence, or rough
air, but rarely are they told why, or what may happen if they don't.

1

In order to clearly see the answer to these questions, let's hypothesize two scenarios. First, assume you are charging through the air well above maneuvering speed. Now suddenly grab the yoke with both hands and give a good healthy tug right to the stop. The aircraft will abruptly pitch up, your butt will be jammed down in the seat, your cheeks will sag down toward your chin, and all that force will bend the wings back, or worse, break 'em off! And as pointed out earlier, airplanes don't fly very well when the wings fall off. Now let's do the same thing while cruising along at or below maneuvering speed. When we abruptly yank the yoke full back, right to the stop, the airplane will respond by zooming up until it runs out of poop (until all the energy is dissipated), and then it will what? Stall, that's what! Now which would you rather do, recover from a stall or attempt to fly an airplane from which the wings just departed? What we've seen graphically illustrated is that maneuvering speed is simply the speed at which an airplane will stall instead of bend or break.

While cruising along in light chop receiving the occasional jolt that tightens the belt across the lap, everyone who operates a modern airplane with an aural stall-warning device has at one time or another heard the stall-warning horn sort of ticking with an occasional beep, beepity, beep. And if it has a visual (red light) stall-warning device, it will flicker in that situation. What this is telling pilots is that they have encountered a vertical gust, an updraft, which, by striking the underside of the wing has momentarily increased the angle of attack above the crucial stall point (see Chapter 3). The airplane has briefly stalled, but the stall was so quick and transitory that the application of recovery technique was unnecessary. The airplane simply flew out of it before the pilot could react. However, if this same event—the vertical gust—should occur at high cruise speed, well above maneuvering speed, the airplane structure might well be damaged and that's why it is so necessary to go to and maintain maneuvering speed when penetrating turbulence.

The fact that maneuvering speed is that speed at or below which an airplane will stall rather than have the structure yield also explains why Va is a higher speed for a heavily loaded airplane than for the same airplane operating substantially below maximum allowable gross weight. An airplane operating at max gross can withstand a harder shove upward on the underside of the wing before it bends or breaks than the same wing carrying a lighter load. Also, an airplane stalls at a higher speed when it is heavy than when it is lightly loaded. This explains why many approved flight manuals and pilot's operating handbooks list several different values for maneuvering speed, depending on the weight at which the airplane is being flown.

Another point to remember is that although the airplane may take the bumps resulting from flying through turbulence at a high rate of speed without bending or breaking, the structure is being weakened to the point that it may ultimately bend or break. This is why there was, a few years ago, a huge AD (airworthiness directive) on the spar in the Cherokee series that was later modified when it was realized that all the spars which had cracked were in airplanes used for pipeline patrol, which necessitated flying at low altitudes where the turbulence is more constant.

## II The importance of weight and balance

Every student pilot is taught how to work a weight and balance problem and is told that it is important to make sure every flight is loaded within the envelope for both takeoff and landing. But does he or she really understand just why this is so and the disastrous effect of being out of the envelope?

In my work as an expert witness in aviation litigation (almost always for the defense of a manufacturer being sued because its airplane failed to save a pilot who blundered into seriously injuring himself or worse), I have had occasion to study literally hundreds of accident reports. And one fact stands out. It actually jumps up and hits you in the face. I'm referring to the fact that in almost every aviation mishap, weight and balance is a factor. It may not be the determining factor, but it is a factor. And when it is the bottom-line cause of the accident, it is almost invariably fatal! The documented cases where this is so are literally appalling. Two with which I am familiar come to mind, both involving the Twin Beech (Model 18).

In one, a group of hunters with their dogs was taking off from a short but adequate field when all the untethered dogs rushed to the rear of the airplane, shifting the CG aft of its allowable limits, causing the airplane to stall, with a disastrous result. In the other instance a bunch of skydivers were sitting on the floor (with the door removed) without seat belts fastened around their thighs. They all crowded to the rear of the airplane, again shifting the center of gravity beyond its aft limit with the same result, a bunch of fatals and a couple of seriously and permanently crippled survivors. In the second case, when the aircraft hit the ground the bodies were thrown up against the pilot and the panel.

A pilot must not only know that it is important to keep the airplane below its maximum allowable gross weight and in the center of gravity envelope, but just why this is so important.

Insofar as gross weight is concerned, admittedly there is a safety factor built in, and any airplane can and will fly perfectly well when it takes off over gross, given enough runway and a low enough density altitude. Landing is a different matter altogether. All airplanes are built to withstand an occasional hard landing, but what if the hard landing is combined with an airplane that is substantially overweight? In this case something is likely to break, or equally bad, the structure is likely to be weakened to the point that it will break at some future time when everything seems to be normal and some other innocent soul is flying it. But even more disastrous than being overweight coupled with the hard landing is the danger of reaching or exceeding the structural integrity of the metal when maneuvering or when turbulence is encountered. In this situation, again hidden damage could result, causing a catastrophe at some future, totally unexpected time.

If an airplane is certified with a maximum gross weight of 6,000 pounds, that's what it weighs sitting on the ground, but take it into the sky and roll it over into a 60-degree bank and it thinks it weighs 12,000 pounds, and if you could put it on a scale that's what it would show, twelve thousand pounds. At its maximum certificated gross weight there is no problem—it is built to withstand virtually any maneuver. But load it up to 8,000 pounds and then make that 60-degree bank or abrupt pull-up (which is really nothing more than a turn in the vertical plane), and suddenly it weighs 16,000 pounds, and it might not be able to take that! Even if it could, there would likely be internal stress damage that will show up later on (Fig. 1-1).

In-flight breakups of aircraft are almost always the result of the pilot losing control momentarily, and as the airplane starts to dive the pilot grabs the yoke with both hands and gives a good healthy yank, overstressing the spar to the point that the wings (or vertical stabilizer and elevator) depart the airplane. And airplanes don't fly very well when the wings fall off (or without the tail, either)!

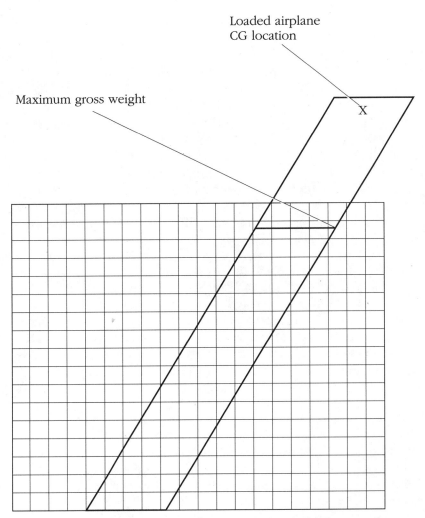

**1-1** *Over maximum gross weight. Danger: Potential structural damage in turbulence and during maneuvers.*

What else can come along and cause the load factor to increase like this and we have no control over it? Turbulence, that's what! If we took that 6,000-pound airplane, loaded it up to 8,000 pounds (remaining within the extended center of gravity envelope) and flew it very carefully, making very shallow climbs, and shallow, gentle banks, we could no doubt get away with it if the air was smooth, but in all the years I've been flying I've never seen a guarantee that I would have smooth air out there. So that's why we won't overload the airplane. But how about staying within the range of the center of gravity envelope?

Pilots are taught to avoid loading the airplane beyond the aft limit of the envelope, but the disastrous results of being out of the envelope forward are largely ignored during training. I vividly remember an accident which occurred at a nearby uncontrolled airport in which a Cessna 206 with the rear seats and rear door removed (placing the center of gravity already forward of its normal location) took off with a load of skydivers sitting on the floor, parachutes at the ready. (Hopefully they had seat belts fastened across their thighs.) Two heavyweights, well over 200 pounds each, occupied the front seats. After climbing up and dropping the jumpers they brought the airplane back to land. On final, when they reduced the power, the nose of the airplane dropped down, and with full flaps and full up trim they lacked sufficient elevator authority to flare for landing.

Surprised at their inability to get the nose up, they added power and went around for another try. Of course, the addition of power and retraction of the flaps permitted them to do this. On the second approach, the same problem reappeared. This time they went ahead and landed anyway. They made a one-point landing, nosewheel first and crumpled up the whole front end of a beautiful airplane! Two very embarrassed fat guys disembarked from the ruins of a perfectly good airplane which had been badly mistreated.

Bad as the situation was, it could have been salvaged by the right-seat occupant relocating himself to the rear for the landing, or even by carrying power throughout the flare and touchdown, in which case if the landing was soft enough the nosewheel and engine mount might have held up instead of breaking and destroying the entire front end of the aircraft (Fig. 1-2). See why it's so important to not exceed the forward limit of the center of gravity envelope?

Being out of the envelope aft is even worse. If the center of gravity is located too far aft of the specified limit, the airplane may lack sufficient elevator authority to recover from an inadvertent stall, in which case an unrecoverable flat spin may be the result. This, of course, results in death (Fig. 1-3)!

The subject of stalls and stall recovery will be treated in Chapter 3. Keep an open mind, and I promise to open your eyes. The problem of the aft CG is aggravated by the fact that having the CG at or beyond the rear limit causes the airplane to become unstable.

CG location

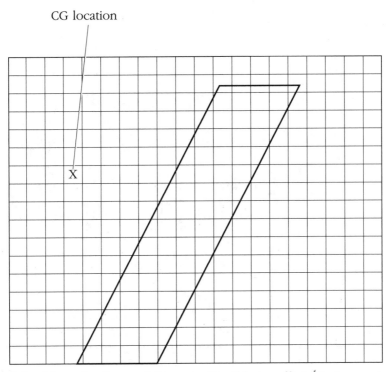

**1-2** *CG too far forward. Danger: Possible insufficient up elevator to flare for landing.*

Has it now become apparent just why it is so important to load your airplane carefully within the envelope, with attention to both the maximum allowable gross weight and the location of the center of gravity? And on the subject of weight and balance, there are several other factors that must be taken into account. Some airplanes have long, skinny fuel tanks or cells located forward of the front main spar. In these models, as you fly and burn off fuel, the center of gravity moves aft, so not only must a takeoff weight and balance be computed, but an anticipated landing weight as well.

Then there's the matter of "zero fuel weight." An airplane in flight has enormous lifting force exerted on the wings, at least the weight of the entire loaded airplane, and more in a climb or turn. To counteract this, we must do something to hold the wings down so they won't fold upward and possibly break off. What we do to accomplish this is create what is called *zero fuel weight.* This is defined as that weight, in pounds, above which any added weight must be fuel in the wings. This becomes especially important with those airplanes

CG location

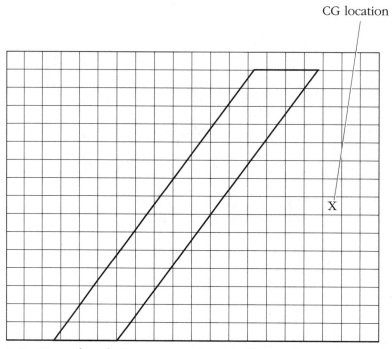

X

**1-3** *CG too far aft. Danger: possible insufficient down elevator to recover from a stall—results in a flat spin—usually fatal.*

which have a fuselage fuel tank, such as the Piper Aerostar. Some Seneca models, because of the great carrying capacity and large cabin, also have a zero fuel weight. This same principle of holding the wings down is why some airplanes with tip (or outboard) tanks recommend using fuel from the inboards first.

Most large airplanes (in addition to a maximum gross takeoff weight, landing weight to protect the structure on landing, and zero fuel weight) also have a maximum ramp weight, based on their estimated fuel burn during taxi and runup.

Now do you see why it is so important to know and understand the weight and balance limitations of the airplane or airplanes you fly? I'm not suggesting that it is necessary to sit down and work out the weight and balance every time you embark on a flight. If you fly the same airplane all the time and are thoroughly familiar with its characteristics and limitations, you can no doubt eyeball your load and know that it falls within the envelope. However, every time you get ready to fly a different make and model airplane, I urge you to sit down and work out a few weight and balance problems at the extremes so you'll have a general idea of the loading characteristics of that particular airplane. Then for the first few times you fly that airplane, using the actual numbers, work out the weight and balance for that particular flight. Be particularly suspicious of rental airplanes that have seen a lot of service. They may have been subjected to hidden stress that is ready to show up when you are operating well within the performance and weight-and-balance envelopes.

Remember, the minute you step outside the performance or weight-and-balance envelopes, you become a test pilot. And I don't know about you, but I'm not paid to be a test pilot, and until I am, I'll continue to fly the way whatever airplane I'm in was designed to fly.

# III Landings

## Touch and goes

I recently received a letter stating, "The touch and go is a timesaving device, not a valid training tool." The writer then goes on to relate several experiences.

> *I have never liked touch and goes, ever since I was a student pilot twenty-some years ago. During my training, the three dumbest mistakes I ever made as a pilot occurred during touch-and-go maneuvers. These consisted in taking off in a fire-breathing Cessna 150 with Full Flaps, taking off in a fire-breathing Cessna 150 with carb heat on, and my dumbest maneuver, taking off in a fire-breathing Cessna 150 with full flaps and carb heat on...Later, after some tail-wheel experience, I gave up touch and goes as an unnecessary hazard. It just isn't worth a groundloop to get one or two more landings per hour. It was then I realized that touch and goes (are) a convenience adopted to save time and lower the cost of training. This is a poor balance of interests.*

He then goes on to relate an incident which serves to reinforce his opinion. An airplane with an instructor and student aboard lost its nosewheel and had a prop strike on a touch-and-go landing. Neither the student nor the instructor were aware of this, and prior to terminating the lesson they flew around for a half hour more. When they finally came in to quit for the day, the landing resulted in substantial damage to the airplane. Fortunately, nobody was hurt, but with an unbalanced prop the result of continuing to fly could have been disastrous.

I started off my answer to this letter by emphatically stating that a landing isn't complete until the airplane has rolled to a full stop on the ground under control of the pilot. I went on to explain, however, that in training students during the presolo landing phase while they are getting the flare down right, the touch and go is a valid technique during dual instruction, but only in tri-gear airplanes, never in taildraggers. The full-stop, taxi-back system does, however, allow the instructor to critique each landing and prepare the student for the next one. While working with a student on the flare, I use touch and goes, but otherwise full-stop taxi backs. If you ever have the opportunity to see one of the real old pros out practicing landings solo (they do a lot of that), you will see a series of full-stop, taxi-back landings.

## Perfect landings

If you want a "greaser" every single time, let me explain just how that can be accomplished. To begin with, a good landing starts with a good pattern. From the time you are on downwind adjacent to the point of touchdown, at which time you make the initial power reduction and commence a glide straight ahead, you keep asking yourself, "Am I high, low, or just right?"

After applying approach flaps, and making another power reduction, you turn onto the base leg. Stabilize your airspeed with the elevator (see Chapter 2), adjust your rate of descent with the throttle, and extend another ten degrees of flap. Keep asking yourself, "Am I high, low, or just right?" And keep glancing from your airspeed indicator to the runway, airspeed, runway, airspeed, runway, as you turn final and extend flaps all the way.

As you get close to the runway, begin to break off the glide and reduce the power to just a tad above idle. At this point, sight down the edge of the runway, and when you are a couple of inches off the ground, maintain a steady back pressure on the elevator. Take the slack out of the stick (or yoke). As the speed bleeds off, the yoke will become loose in your hand. Do not permit it to do so. Your objective at this time is to hold the airplane off the ground as long as you possibly can. If you follow this procedure, you'll find yourself rolling down the runway wondering just when you touched down. Do this, and you'll grease it on every time!

For a discussion of stalls, see Chapter 3, but for now you should know that when an airplane is in ground effect, at a height above the ground of less than its own wingspan, a full-breaking stall simply can't happen. This is because of the change in the downwash on the tail surface (horizontal stabilizer or stabilator). So it is impossible to smash down nose first if you are that low. Of course, you can still drop it in and get a hard landing if you really try hard.

# IV  Propeller safety

For starters, let me point out that I'm terrified of props. I'm paranoid about the subject, and I admit it. That's not to say I won't prop an airplane to start it because I will do that, but I work on the assumption that every prop in the world is alive and is out to get me. I've seen what a propeller can do to the human body, and it ain't pretty! I personally know two one-armed CFIs who donated arms to props, plus I know of another serious injury and one fatality.

Although not a huge factor in the total aviation accident picture, propeller mishaps continue to occur, and they are so easily preventable there is simply no excuse for any prop accident, ever. Over a recent twelve-year period, there were an average of eleven prop strike accidents per year, with a high of twenty-one and a low of five.

Propellers contacting and damaging some part of the human anatomy are no doubt the most serious aviation accidents which occur on the ground. Such contact is almost invariably fatal, and if not fatal the loss of a limb or other extremely serious injury is the likely result. The usual prop accident occurs when passengers (or crew) are deplaning with the engine(s) running. The prevention here is very simple: never permit anyone to leave an airplane until the prop has completely stopped rotating. This also applies to boarding an airplane. A recent study of some 208 airplane prop strike accidents determined that 48 percent of them were caused by passengers boarding or leaving the airplane. This is by far the largest group of people involved in such incidents.

Not only should no one be permitted to enter or leave an airplane unless the propeller(s) is (are) completely stopped, but passengers should always be briefed regarding the dangers of propellers and advised to remain clear at all times, even when stationary. Bystanders are also subject to attempting to occupy the same space as propellers, and we all know it is physically impossible for two objects to occupy the same space at the same time. In fact, the second most common kind of individuals involved are ramp, ground personnel, and other pedestrians who walk into these invisible whirling meat cleavers. This group accounts for 25 percent of the total. It's the pilot's duty to make sure the area is clear prior to engaging the starter, and shouting "Clear the prop!" is not sufficient. Pilots must actually look around the entire area to personally assure themselves that it is indeed clear. Engine start-up is so routine a matter that many pilots become complacent to the point that they bury their heads in the cockpit while yelling, "Clear!" without ever looking around to see if the area around the airplane is indeed clear of living things (or inanimate objects).

The FAA recommends that the rotating beacon or flashing strobe be on any time the engine is running, on the ground and in the air. What they don't say but should is that such visual warning ought to be turned on before the engine is started. In addition to the verbal shout, "Clear the prop!," this helps to alert any bystanders that you are about to fire up a dangerous machine. On a busy airport with quite a bit of ground traffic, the noise level is likely to be such that an oral warning might not be heard, so the visual backup provided by the flashing beacon could well be the only real warning a pedestrian receives. A rotating propeller itself makes a poor visual target. It is nearly invisible as it whirls around, even at idlespeed. On the other hand, a flashing or rotating beacon is designed to attract the eye like a magnet. A good technique for being sure you don't leave the master switch on is to leave the rotating beacon on all the time. Then after shutdown, as you walk away from the airplane, if the master is still on, the rotating beacon is likely to grab your attention and alert you to the situation.

And while we're on the subject of engine runup, I'm willing to wager that your instructor never taught you the second reason why it is important to face the airplane into the wind when you do your pretakeoff runup. Everybody is taught to point into the wind for better engine cooling, but equally important and often neglected is the fact that a crosswind or, worse, a tailwind puts quite a strain on the prop itself. This is particularly true of adjustable (constant speed) props. Didn't know that, did you?

The next most common form of propeller-meeting-human-being accidents are those which occur when an individual is attempting to hand prop an airplane to get it started. This group accounts for 15 percent of the total. Hand cranking a prop is not particularly hazardous if it is done properly. Remember, not too many years ago that was the way most general-aviation airplanes were routinely started. Now it is usually only necessary when the battery is dead and there is no auxiliary battery cart and cable available. Even so, it is not a good idea to attempt to start a high-compression engine by hand propping, unless you have a desire to strain your back.

Countless damage accidents, and not a few injury accidents, have resulted from someone hand propping an airplane without a competent person at the controls. No one should attempt to start an airplane this way, but if it must be done, the airplane should be securely anchored (tied down) and the throttle barely cracked. I am personally aware of two cases where airplanes were hand propped with nobody at the controls and the throttle wide open, both resulting in substantial damage. The FAA has a great movie clip on this subject. Be sure to see it if you get a chance. The safety  program manager at your local FSDO (Flight Standards District Office) either has a copy or can get it for you.

It is my belief that every pilot who operates light airplanes should be properly trained in hand propping techniques and procedures, and the odds are most modern pilots have not been taught the proper technique for doing so. This is a classic example of something your instructor never taught you, probably because they never learned it themselves. This is a procedure every pilot should know, for one never knows when it might become necessary. I'll never forget a story told by the then manager of our local GADO (before it became a FSDO—Flight Standards District Office—by being combined with an ACDO—Air Carrier District Office—it was a GADO—General Aviation District Office) at an examiner meeting several years ago. Seems that when he was a civilian pilot examiner, a young lady came in for a private pilot flight test with a J3 Cub. When they completed the oral and got ready to fly, she boarded the airplane and strapped herself in. He asked her how she intended to start the airplane if she didn't intend to prop the airplane to get the engine going, and she replied that her instructor had always done that for her and had never showed her how. He sent her back with a pink slip and with instructions to her CFI to teach her how to start an airplane! I wonder what she expected to do if she found herself in a situation requiring her to fly somewhere and there was nobody around to prop the airplane for her. (Fig. 1-4.)

If you haven't been taught the technique, here is a step-by-step description of the procedure:

1. Always assume the prop is hot (alive).

2. Make sure the airplane is on a good, firm, level surface so you will have good traction. You don't want your feet to slide out from under you as you fall into the spinning propeller.

3. Always assume the prop is alive.

4. Be sure there is a qualified person aboard and at the controls. Most hand-propping accidents occur when a pilot attempts to start an unattended airplane, or one with an unqualified individual seated inside.

5. Always assume the prop is alive.

6. Have the ignition key (if any) in your hand, or on top of the panel in plain sight while you do your walk-around inspection. If there is a switch and no key, ascertain that it is in the "OFF" position. Do this yourself. Trust no one!

7. Always assume the prop is alive.

**1-4** *Harold Thompson props a J3 Cub. Notice how he is ready to step back when the engine fires.*

8. With the switch OFF, grasp the propeller near the hub and pull straight forward as hard as you can to confirm that the airplane is firmly anchored.

9. Always assume the prop is alive. (Are you beginning to get a message here?)

10. With the switch OFF, pull the prop through (forward) until the top of the compression stroke is reached.

11. Always assume the prop is alive.

12. Call for switch ON and receive acknowledgment from the responsible person in the cockpit. Now you **know** it's alive!

13. Place both hands flat against the blade (do not curl your fingertips around the propeller blade) and sharply snap the prop through as you step back.

14. If the engine fails to start, reset (with the switch OFF) and try again.

15. Always assume the prop is alive.

16. If the engine still won't start, with the switch OFF (and confirmed), pull the prop through several times and try again.

17. Always assume the prop is alive!

Remember, prop blast blows sand and gravel. Always be considerate of others when performing a runup. Propellers create missile hazards in areas where loose gravel, sand, dirt, and other small objects can be blown about. Be sure to check the area you are taxiing into or away from to ensure that your prop blast won't cause injury or damage, and use only the minimum power necessary. Aircraft ground service personnel should be wearing appropriate eye and ear protection, but bystanders are at risk unless kept at a safe distance.

I don't think I'll ever forget the time a renter pilot had taken one of our school airplanes on a 1,200-mile trip. On the way home he stopped for fuel at an airport some 500 miles from where we're located. When he started up to complete his trip, he blew dust and sand back onto a group of spectators who were standing behind the airplane. Unknown to the pilot, another one of our customers witnessed the event and reported it to me. When the guilty guy got in, I counselled him regarding what he had done, and he said, "My goodness, Howard, have you got spies everywhere?"

My reply was, "Of course, I do! Don't think for a minute that you can get away with anything without your Old Dad knowing about it."

Finally, if you have ever observed a spinning propeller over a puddle of water, you know how the rotation of the prop can pick objects up and draw them into itself. It is just as easy for a propeller to pick up small stones and gravel and draw them in, causing nicks in the propeller itself. Unless treated, this condition can cause blade fracture, and when this happens in flight it is likely to ruin your whole day, week, month, or whatever.

We must be particularly careful to be watchful when children or dogs are walking around on the ramp. And here's a tip for you. Whenever you shut down and park an airplane, whether you are securing it for the day or going out again, leave the rotating beacon or tail strobe on. This does two things. One, it ensures that it will come on when you next turn on the master switch, even before you fire up the engine. And two, it will prevent you from inadvertently leaving the master switch on when you park the aircraft because you will be sure to notice the beacon or strobe as you secure the airplane or walk away from where you parked it.

# V The sod strip

I can't imagine why, but many flight schools have a rule prohibiting their students and renter pilots from taking their airplanes into and out of sod strips. I know they sometimes blame this rule of their own on their insurance carrier. I am familiar with no insurance policy that specifically prohibits the use of sod strips by any aircraft covered. Thus it would seem that such a rule is personal to the operator. Of course, every operator can impose any rule, but if the operator forbids students and renters from taking equipment into grass fields, the operator is depriving them of the opportunity to gain desirable experience, and denying them the opportunity to go to a great many desirable places that are served only by grass airports. Besides, sod strip operations are good practice in the unlikely event you should have to make a forced landing.

As a pilot examiner, when giving a practical test for the Private Pilot Certificate, in testing for Area of Operation IV, Task C (diversion while on a cross-country flight), I would occasionally hypothesize worsening weather and divert the applicant to an airport devoid of pavement. I would always make sure I was directing us to an airport with good, firm sod strips with good drainage and no soft, muddy places, and one which contained no hidden obstacles. Even so, more often than not the applicant, upon arriving over the airport, would exclaim, "I can't land there! There are no paved runways."

My answer would be, "That's an airport, and if this were a real emergency requiring a diversion, you'd better be prepared to land. Go ahead and try it."

This might get me a response to the effect that the school whose airplane we were in didn't allow it, in which case we would forgo the landing, and when we finished the checkride I would counsel, usually unsuccessfully, the operator of the flight school.

However, if I received no such response, the applicant would then proceed to make a normal routine landing, and as we were rolling out after touchdown would turn to me and say, something like, "Hey, there's nothing wrong with this. It's just like any other landing!" Surprise, surprise!

A pilot who is not aware of the fact that a grass strip, if it has a good, firm surface, can be just as good a landing area as a paved runway is missing out on a great many places he could otherwise go. And one usually doesn't even have to use soft-field technique, although most uninformed pilots believe such a procedure to be necessary when landing on any sod surface.

If you have never landed on a sod strip, by all means get yourself an instructor and go out and do it. You don't even need an instructor. Just be sure to be alert, look around, and fly a proper pattern. When landing on wet or frost-covered grass, brakes should be used sparingly if at all. It is better to let the airplane roll out after touchdown at any time on any surface, but it is essential on a wet or icy surface, whether paved or not. If the field is unattended, it is a good idea to drag the landing area first before attempting to land. Make a low fly-by as slowly as consistent with safety and look the field over carefully for ruts, obstructions, and uneven surface areas. As always, in aviation the basic rule applies. There should be no surprises.

There are some situations where it is better to fly from grass fields. A couple of years ago a customer of mine who owns and operates a pressurized twin in his business bought a beautifully restored 1942 Boeing Stearman (a PT17). And I was fortunate enough to be asked to check him out in it. We are both based at a busy tower-controlled airport with eight paved runways (four strips of pavement). About 30 miles down the road is a small field with both north-south and east-west turf runways. The approaches for three of these sod runways are clear (there are wires along the road at the south end of the airport), and the drainage is excellent, so that's where we went to get the guy checked out in his nice new airplane.

Despite the fact that Mel is a perfectly competent pilot both on instruments and when visual in his pressurized twin, it took over twenty hours of instruction before he was capable of taking off and landing the Stearman off the grass before I could introduce him to operations on the pavement, but this was a taildragger problem rather than a grass problem (see Chapter 10). Of course, during that 20-odd hours each time we went out to fly, we took off from pavement at our home base and landed on the concrete when we returned, but I had to assist on both the takeoffs and landings for that entire period. He finally reached the point that he could successfully manipulate his airplane on and off the pavement, but barely so.

Some airplanes just prefer grass. That's all there is to it. Most tail-draggers do better on sod, particularly those with close-coupled gear, such as the Stearman and the Luscome.

Another often-overlooked advantage to having experience on sod fields is the fact that it is great practice for the time when you might have to make an off-airport landing. You know, one of those dreaded "forced landings." Back when I was trained, a good deal more attention was placed on practicing forced landings than is done today. This is because today's engines are infinitely more dependable than the engines were back then. But even so, engines do fail. It is usually the fault of the pilot, since fuel mismanagement (either fuel exhaustion or selection of the wrong tank) is the most common cause. However, whatever the cause, mechanical failure or pilot stupidity, it does happen, and when it does it need not be a big deal at all unless the pilot makes it one. (For a full discussion of the subject, see Chapter 3.)

# VI  Night flying

Although the current experience requirements for private pilot certification include 3 hours of night dual instruction, including ten landings at night, many aviators who achieve the private certificate are not really ready to command an airplane at night, either alone or with one or more passengers. In fact, I'm not sure the 5 hours and 10 landings at night as sole manipulator of the controls required of commercial applicants adequately prepares an individual for night operations. Its a completely different world out there at night.

As demonstrated by the accident statistics, night flying involves a much higher degree of risk for both VFR and IFR flying than do day operations. And although the concern of most pilots regarding flying at night is the possibility of loss of power and the subsequent difficulty of effecting a safe emergency landing, with today's dependable engines this is in reality only a minor concern. Very few night accidents are the result of engine failure. The primary cause of nighttime accidents is spatial misperception on the part of the pilot.

The only way to acquire competence in night operations is through experience. This would seem to bring us back to the perception of a deficiency in training for flying at night. If you are like most pilots, you don't fly very frequently at night, and to maintain currency for carrying passengers at night you will from time to time go out on a clear night and do the three landings. I suggest that you would be well advised to do the three landings at three different airports. There's nothing more peaceful than flying from one point to another on a clear night with the moon and stars out. The air is usually smoother, and VFR navigation by pilotage is a cinch. The cities and towns are laid out like jewels on black velvet.

The two airports away from home at which you land need not be particularly far apart, but this technique of doing your landings at different airports gives you the opportunity to experience the differences in visual perceptions that occur between flying in the daytime and flying at night. And on the subject of night currency and currency in category and class, let me tell you about Jim. Jim (now retired) was the chief pilot for a large multinational corporation with a fleet of large airplanes. He lived on the flight deck of a Grumman GII. He had a daughter in college some 350 miles away, and from time to time he would rent one of our single engine airplanes and take his wife to visit their daughter. Each time, if 90 days had passed between such trips, he would go out by himself and do three takeoffs and landings before boarding his wife and taking off to see their daughter. If he can do this, so can we. And by the way, you should see that guy do a walkaround preflight of the airplane. He inspects it like the finance company coming to repossess it! I wish all my students could watch him. They'd get to see just how a real professional does it.

# VII  VFR cross-country

When I was trained in the early 1940s, there were four kinds of navigation in use: pilotage (flying by reference to ground objects), ded reckoning (calculating time and distance), celestial navigation (by reference to the heavenly bodies—and we don't mean the gorgeous human female form), and radio navigation (flying the "beam"—one of four courses put out by the Adcock Radio Range). By far the most common was pilotage. Back in the days before the advent of VORs, we practically navigated from treetop to treetop. And this kind of true VFR navigation is rapidly becoming a lost art.

The only pilots who still know how to find their way by pilotage are the old guys like me who remember how we used to do it. I know a fellow who had a nice airplane with no less than eleven radios in it, and when one of them failed, he left the airplane some 300 miles from home and took a carrier because he feared he couldn't get home without all his navigation equipment. And on another occasion, I recall a guy who landed at the old Chagrin Falls airport (east of Cleveland) and asked where he was and how to find Cleveland Municipal Airport (now Cleveland Hopkins). He had flown all the way from Louisville by pilotage and ended up over 40 miles from where he thought he was. And I am aware of several more such aviation adventures, including the one in which a pilot landed at what he thought was his destination (Pellston, Michigan, an uncontrolled field), and when he taxied up to the terminal he was shocked to see a sign proclaiming, "Welcome to Traverse City," a tower-controlled airport more than 30 miles from Pellston.

I recently heard of a case in which an applicant showed up for a private pilot practical test with an airplane equipped with a Global Positioning System receiver. His entire cross-country plan consisted of looking up the latitude and longitude of the destination airport, expecting to plug it in as his GPS waypoint and fly directly there. The examiner accepted this, and when they got to the airplane, after completing the walkaround preflight inspection and declaring the airplane airworthy, the applicant set up his destination waypoint in the GPS, and they took off. After traveling some 15 miles or so, the examiner turned off the Global Positioning System receiver and asked the applicant to proceed without it. The fellow instantly became hopelessly lost! Of course, the examiner had no choice but to bust the applicant.

Not only are such occurrences humorous (they really generate quite a bit of laughter), but they are tragic in their implications. Any equipment, and particularly avionic equipment, is subject to failure. And the interesting thing is, this stuff fails at the most inopportune times.

Most pilots demonstrate their ability to plan and fly a cross-country trip by pilotage on their private checkride and thereafter promptly disregard what they have learned. Very few ever again draw a course line on a chart, complete a trip log, marking checkpoints every few miles, and carefully noting the time as each checkpoint is passed. Part of the fun of traveling is enjoying the sights as we cross the countryside, and if your VFR navigation skills are not up to par, you are likely to miss out on a good portion of this experience.

I recall reading an article several years ago by Leighton Collins in which he related the following story:

> *"A flight instructor encountered a former student of his on the ramp at Teterboro, New Jersey. The guy had sectional charts spread over the wing of his low-wing airplane and trailing on to the ground, and he was struggling with a yardstick to draw a course line from one chart to another. The instructor asked his former student what he was doing and the fellow answered that the was planning a trip to Florida. On hearing this, the instructor folded up all the charts and pointing east said, "Takeoff, point the airplane that way, and when you come to the first ocean, turn right! Eventually you'll get to Florida."*

Of course, this is a gross exaggeration, but basically that's more or less how it was done in those days. And you'd better know how to do it today if you expect dependability from your VFR cross-country flying because there will surely come a time when you won't have all the goodies you are used to.

Please don't misunderstand me. I'm not saying that it was better in "the good old days," but it sure was different, and I think it was more fun to be an aviator. One didn't need to know as much, but he/she had to be more skillful at manipulating an airplane around in the sky and arriving at his or her destination without the aid of all those electronic marvels.

A flight instructor recently told me that he believes a majority of the country's VFR-type private pilots would become hopelessly lost if they were forced to navigate by pilotage, and he just may be right. I personally know an instrument-rated pilot who belongs to a group that flies out to a distant airport every Saturday for lunch. This group is based in Southern Lower Michigan, and on those occasions in which the selected location for the luncheon expedition is in another state, the guy stays home. He knows that if it's in Michigan and the destination is east, west, or north and he goes too far, sooner or later he will come to a large body of water, but if he has to go south and cross a state line, he won't know where he is! And this guy is an instrument-rated pilot!

The old pioneers who surveyed our country did VFR pilots an enormous favor. They laid out the country in a square grid. Except for an occasional diagonal, usually caused by terrain features such as hills and rivers, all the lines (section lines and roads) run north-south and east-west. With practice, you can eyeball the angle and line it up with your course line on the sectional chart.

But remember, you're rarely going right where you're pointing. The wind must be taken into account! This is easy if you're low enough for the drift to be apparent. Just eyeball the ground, see what the drift is, and adjust accordingly. However, if you're too high for the drift to be readily apparent, you must look for other clues as to what's happening, and there are lots of 'em available. Smoke, streaks on water, and your shadow moving across the ground are but a few. Of course, the wind at altitude is never just what it is at the surface, so you must adjust for that difference as well. And with today's equipment (Loran C and GPS), it is even easier to adjust for the wind by comparing your course with your track. In any event, a pilot must be constantly aware of what the wind is doing to him at all times.

Once again, the fun and challenge of flying VFR by pilotage without the aid of any avionic navigation is far from the only benefit to be derived from this activity. One of the joys of flying VFR is to be able to look around outside as the country passes beneath us and spot the interesting sights our great country has to offer. This factor should definitely not be ignored.

# 2

# Control considerations

## I can't fly

I can't fly, you can't fly, only birds and insects can fly! (Fig. 2-1). However, airplanes can and do fly. In fact, they do a beautiful job of it if we pilots just leave them alone and let them do it. The only thing the airplane can't do for itself is think. That's our job. Just consider, the designer creates a wonderful machine, the manufacturer painstakingly builds it, and then they hand it to us, the pilots, to use it or abuse it as the case may be. The really good pilots are physically lazy but mentally alert. The best pilots are the ones who do the least. They let the airplane do the work while they do the thinking.

One of the most difficult concepts for the flight instructor to get across to the beginning student is the fact that the airplane is inherently stable. If left alone, it won't fall out of the sky. And one does not have to be constantly driving it as if it were an automobile. We need only point it in the direction we want it to go, trim it out, and let it take us to our destination. As every pilot knows, flying straight and level consists of constantly making a series of small adjustments. Some people refer to them as corrections, while others of us prefer to call them what they really are — adjustments.

Cross-country flying by pilotage is rapidly becoming a lost art. We now have so many marvelous avionic assistants (read "toys") to help us with our navigation, or in effect do it for us, that we are no longer required to select a landmark along our course and fly toward it until we can pick out another one farther along and so on. This is only part of the difference in the way we fly today as opposed to the way we did it only a few years ago. As recently as the late 1940s and even through the mid-1950s, general-aviation flying was quite different

**2-1** *Birds, insects, and some mammals can fly, but not human beings.*

from what it is today. I'm not saying it was better in the "good old days," but it sure was a lot more fun!

Today's pilots don't seem to have any idea what that rudder is all about. Those pedals are just there to get in the way of the pilot's feet, or perhaps to provide a footrest. Many pilots today simply plant their feet firmly on the floor and drive the airplane through the sky as if it were an automobile, treating the yoke as if it were a steering wheel. And they can do this more or less successfully because the designers have made the airplanes so much better. For small heading changes with a very shallow bank, it can be accomplished and the pilot can get away with it. However, if we observe an old timer at the controls, we see something entirely different. This pilot, who was trained in older airplanes, carefully trims the airplane to fly straight and level, places his toes lightly on the rudder pedals, folds his arms across his chest, and goes about his business. His hands are free to fold (or unfold) his charts, drink his coffee, eat his sandwich, or whatever else he wants or needs to do while looking around the sky for traffic. If a minor heading adjustment is required, this guy merely exerts a small amount of pressure on the rudder pedal in the desired direction, the airplane rolls with a slight bank, and voila!, the airplane is now pointed in the correct direction. If a slight altitude adjustment is required, the old-time pilot applies a small amount of pressure (either up or down, fore or aft) to the elevator control (either yoke or stick), changing nothing else (neither trim nor power) until the desired altitude is again reached, and again he lets go and lets the airplane take him where he wants to go, perhaps with a very slight amount of retrimming.

This technique drastically reduces the pilots' workloads, freeing them up to divide their attention and scan the skies for traffic. And this business of looking around outside the aircraft is another area the modern pilot is prone to ignore, or at least to slight. There are so many wonderful toys in the cockpit and on the panel that make demands on the pilot's attention that there is little time for surveillance of the area outside for other traffic. Just a few years ago, almost the entire emphasis in primary flight training was on coordination of eye, hand, and foot, and on looking around outside. We were exhorted to "keep our heads on a swivel," and if we failed to do so we were told that our heads were in a most unpleasant place, not to mention the physical impossibility of getting it there! Today, with the improvement in aircraft design rendering manipulation of the airplane easier, coupled with the sophistication of the airspace and equipment making demands on our attention, it's no wonder that the pilot is prone to neglect the rudder and to fail to exercise proper surveillance. And with the speed of the modern airplane resulting in enormous closure rates, it is even more important than ever before to be alert for traffic. In the en-route airspace the skies may not be crowded at all, but in the terminal areas as we approach an airport at a common altitude, it can get pretty hairy.

Another thing old-time pilots do differently: they don't fly "by the numbers," but rather they are more likely to fly by feel and "sight picture" when operating a fixed-pitch, fixed-gear airplane VFR in VMC conditions. To take off, he or she lines the airplane up with the centerline of the runway, applies power, comes back on the yoke or stick until the top of the cowl lines up with the horizon, and then waits. When the airplane is ready, it flies itself off. The modern technique is, of course; to hold the airplane down as it pounds down the runway until the magic number comes up on the airspeed indicator, and then rotate! This kind of flying by the numbers is required when operating a high-performance, complex single or a twin, but the old-timer's method works just fine with the simple airplane. Pilots trained the modern way are likely to panic if they lose their airspeed indicator, as numerous experienced and low-time pilots have done, to their detriment. What enables the old timers to keep their cool when they lose their airspeed indicators is the certain knowledge that whether they are VFR or IFR, if the aircraft's attitude is right and the power setting is right, the airspeed must be in the ballpark. Remember the simple formula that always works: attitude plus power equals performance.

A few years ago a fatal accident resulted from the failure of an instructor to impress on his student the fact that if the aircraft attitude is right, and the power setting is right, the airspeed just has to be within tolerable limits. Here's what happened: A pilot with a fresh instrument rating on his pilot certificate while flying in IMC over the Rocky Mountains noticed his airspeed indicator showing a gradual reduction. The pilot notified ATC that he was losing airspeed. In point of fact he was not slowing down at all. His pitot tube was simply icing up. In a desperate effort to increase his speed and avoid a stall and possible loss of aircraft control, the pilot pitched the airplane down and dove right into the side of a mountain ridge. The engine was operating at near full power, and all aboard were killed. All this because that pilot was totally dependent on his airspeed indicator. And of course because he wasn't knowledgeable enough to realize what was really happening. No doubt denial had taken charge. He simply didn't believe it could be happening to him.

Then there's the fairly recent case of the air carrier jet, full of cash-paying passengers, which also developed a false reading in its airspeed indicators (plural) because of ice forming in the pitot tubes. Even pitot heat is occasionally inadequate to prevent the formation of ice in the pitot system. As the ice builds up, the first false reading is a higher than actual speed. Then as the tube begins to completely close up, the apparent speed falls off to a much lower than actual speed until finally the needle on the airspeed indicator goes to zero and just rests there as if the airplane were sitting still on the ground. The air carrier pilots (plural again) must have caught the change at the beginning of this process, for they thought the airplane was going too fast, so they reduced power and pulled the nose up to slow down, which, of course, is what happened. In fact the airplane slowed down to the extent that it stalled and entered a spin! Just the opposite of the Bonanza pilot who dove into the mountainside.

In the case of the pilot with the brand new instrument rating on his pilot certificate, it is possible to understand why he did what he did (we can blame his instructor, his lack of knowledge, his inexperience, or any number of other factors), but it is literally incomprehensible that no less than three experienced air carrier pilots would all suffer brain fade at the same time and do what they did. What could they have been thinking? The answer is that they weren't thinking. They were reacting as they had been taught, based on reliance on the airspeed indicator.

In the days before attitude instrument flying as it is currently practiced, instrument flying consisted of keeping the airplane upright using the Stark system (1,2,3—needle, ball, and airspeed) while following a colored airway (one leg or another of the four-course Adcock Radio Range). And the airspeed was the least of these. It was the only one which had a backup—the altimeter. The pilot centered the needle in the turn and bank indicator (later called the slip and skid indicator, and finally the turn coordinator) with the rudder, centered the ball with the stick (what we now know as the yoke), and checked the airspeed indicator for pitch! How does that sound to those of you who have been taught to "step on the ball" to control a slip or skid? This Stark system works under most, but not all, flight conditions, and we've come a long way since the days of the Stark system, what with flight directors, horizontal situation indicators, and all the other marvelous new stuff we have today. The current method of attitude instrument flying always works in every flight situation.

However, when we add the sophistication of today's airspace to the improvement in aircraft design and the new devices and techniques, it is no wonder that today's pilots are prone to neglect the rudder and fail to adequately scan the skies around them as they proceed along their way. Basic airplane manipulation has become easier, and flying as a means of transportation has become infinitely more efficient, but we have paid a price. What we must watch out for is to not permit the price to become so great that we end up with a net loss, rather than a gain. If we keep in mind that we can't fly, but airplanes can, and we are still responsible for observing and avoiding traffic, perhaps there will be no price to pay at all, and only benefit will result from the progress we are seeing. Maybe there is a free lunch after all. But to collect this bonus, we must all do our part as well.

# Some fundamentals of aircraft control
## I Pitch to airspeed, power to altitude, power to airspeed, pitch to altitude?

Several years ago a designated pilot examiner would ask private applicants one question by way of oral quiz, and only one. The question was this: "Which control makes the airplane go up?" If the applicant answered, "The throttle makes the airplane go up," the applicant passed, but if the applicant said the elevator makes the airplane go up, the applicant failed the oral and thus failed the entire flight test!

Sooner or later every student pilot becomes aware of the age-old controversy over which controls what, and sides are chosen as the argument rages. Flight instructors often demonstrate one theory or the other, depending on which opinion the individual instructor favors.

The demonstration of power to airspeed goes like this: The instructor says to the student, "You think the elevator controls airspeed? Come on, we'll see." They then taxi out and line up on the runway, and when they are cleared for takeoff the instructor starts to furiously pump the yoke (or stick) back and forth, whereupon the student asks, "What on earth are you doing?" And the student gets this answer, "I'm trying to get up enough airspeed to take off!"

On the other hand, if the instructor favors the power-to-altitude theory, the instructor will say to the student, "You think the elevator controls altitude? Come on, let's see." They then take off, climb to a nice, safe altitude, clear the area with at least a 90-degree turn in each direction while looking carefully around, and the instructor retards the throttle to idle, turns to the student and says, "You think the elevator makes the airplane go up? Okay, pull back on the yoke (or stick) and make it go up!"

The fallacy in both of these arguments is, of course, that there is an overlap in the function of these two controls, and each controls altitude or airspeed, depending on the flight condition. Currently, the favored approach to flight training favors pitching to altitude and powering to airspeed, but the downside of this system requires the student to juggle two variables instead of just one during the landing approach. The rationale behind teaching primary students to pitch to altitude and power to airspeed is that later on when the student undertakes instrument instruction, or checks out in a high-performance single or a twin, this is the way it will be done. The assumption is that the student pilot lacks the ability to make the transition from powering to altitude and pitching to airspeed when the time comes. It is just another example of the FAA insulting the intelligence of student pilots. It is much easier to juggle one variable than two, and perfecting the student pilot's technique in making a stabilized VFR approach in the pattern reduces the work load (not to mention the guesswork) drastically if the pilot establishes a stable airspeed with pitch and adjusts his descent with the reduction or addition of power as required to end up right at the desired point on the runway, running out of airspeed and altitude just as the main gear wheels kiss the pavement.

Think of it this way: You are driving along in your automobile and you come to a hill. Without changing the amount of pressure you are exerting on the accelerator pedal, you start up the hill. What happens? Let's see a show of hands. You in the front row, that's right, you slow down. Now you crest the top of the hill and start down the other side. If you still maintain the same pressure on the gas pedal, you'll speed up. An airplane works the same way. If you pull back on the elevator control, the airplane will pitch up and start to slow down, and if you shove down on the elevator control, the nose of the airplane will pitch down, and the result will be an increase in speed. Therefore it is the pitch attitude of the airplane that determines its speed. If you hold the nose of the airplane slightly down and add power, the airplane will go faster. This is because the addition of power makes the airplane want to climb, but since you are holding it down to prevent a climb, and since the nose attitude is slightly down, the airplane has to speed up. Thus power affects speed as well. Both the ability to climb and the ability to go faster are dependent on the power available from the engine over that being used at any given time.

# II  Why must I learn that?

Over the years every flight instructor in the world has heard students complain, "Why must I learn to do that? I'll never do such a thing once I have the certificate."

With respect to ground reference maneuvers, it is not difficult to relate them to the concept of wind correction for the student. But relating such commercial maneuvers as chandelles and lazy eights to the necessity for smoothness on the controls gets a bit more difficult. And when it gets to steep turns, the student must be made to understand how load factors are increased and vertical lift is lost. Then there's the matter of division of attention.

A few years ago there was a nudist colony right in the middle of the practice area for our airport. We called it "bare intersection." On warm summer days, students by the dozen would congregate around good old bare intersection and make steep circles around the volleyball court while they watched the nudists play volleyball. It amazes me that there weren't a bunch of midair collisions with all that traffic and everybody with their eyes glued to the ground instead of looking around for traffic.

Holding altitude in a steep turn requires the addition of power and enough back pressure on the yoke or stick to compensate for the loss of vertical lift and the apparent increase of weight. And if you're not careful, it's easy in a steep turn to blunder into an inadvertent stall-spin situation (see Chapter 3).

Students also frequently complain when their instructors require that they always attempt to land right on the centerline of the runway. If you want to know why that's important, come on over and fly with me. A few miles from the airport where I'm based, there is a small, country-store-type airport with a single paved strip that is so narrow that a Cessna 150/152 with its nosewheel on the centerline has both wings extending some 3 feet beyond the pavement on each side. If you land there and don't stay right on the centerline, you're likely to have one of your main wheels off in the mud. At our home field the main runway is 300 feet wide; one can almost land on it sideways. When we take our students over to that small airport, we frequently hear, "My gosh! I can't land there! That's a sidewalk down there, not a runway!"

## III The subject is trim

Did you ever notice how when two pilots are occupying the front seats of an airplane and control is handed from one to the other, the first thing the one taking control does is reach out and adjust the trim? It happens every single time, whether they are VFR or IFR. Pilots do what is most comfortable for them. Some prefer to hold a slight pressure on the yoke, either up or down, while others want to stabilize the aircraft to fly hands-off and maintain altitude, but whatever their preference is, it is not what the other pilot has been doing.

Trim stability makes the airplane seek its trimmed airspeed. Therefore, what we're really doing is trimming for an airspeed, and this can easily be demonstrated, either with an instructor or alone. At altitude, simply adjust the pitch trim for hands-off flight in cruise configuration, at cruise airspeed and cruise power, and leave it there throughout the exercise. Now reduce the power to idle. The airplane will pitch down, speed up, pitch up, slow down, pitch down, and speed up, repeating the process through several oscillations until it stabilizes at the original airspeed, but now it is in a glide.

Next, apply climb power, using whatever amount of right rudder is necessary to counteract the P factor and torque, but keep your hands off the yoke to avoid subconsciously adjusting the pitch. The airplane will pitch up, slow down, level off or pitch down and speed up, pitch up and slow down, pitch down and speed up, and repeat this process until it once again stabilizes itself at the original airspeed, but this time in a climb. Whatever you do to disturb its equilibrium, so long as the trim and the power are not changed, the airplane will seek out the speed for which it has been trimmed.

Of course, if you are only flying a Cessna 150/152 Heavy (or any other extremely low-powered airplane), you can just leave it in neutral trim and hardly notice the difference, no matter what you're doing. But when you move up to a 172 (and especially a 182 or a Cherokee 180 or Archer) all at once you'll discover that you have to trim, or you'll wear out your biceps trying to maintain the desired pitch attitude. I know, I just said we trip for airspeed, not attitude, but a by-product of changing the trim is a change in attitude (which is what results in a change in airspeed).

The majority of pilots go about setting their trim wrong, and it is not because they weren't taught better, but because they are lazy. Next time you are flying with someone else at the controls, watch and see what he/she does. Or better yet, check yourself on how you do it. If you are like most other general-aviation pilots, after climbing to your cruising altitude, you simultaneously pull back on the power and start rolling the elevator trim wheel forward. When it seems about right, you stop. And after several adjustments you finally get it stabilized. You were taught better, but you think this is the easy way.

Wrong! Using this technique you may never get it quite right, and you may find yourself making slight adjustments throughout the entire flight.

Do it right, the way you were taught, and you only have to do it once, twice at the most. After reaching your cruise altitude, shove the yoke forward to the desired attitude and hold it there. Now trim off the pressure you have been holding. Finally, after several seconds while the airplane finds its cruise speed, retard the power to the desired setting. You probably won't have to retrim at all. The altitude and airspeed will be nailed right where you want them and your flying will be much more precise.

By the way, have you ever seen anybody adjust the rudder trim on those airplanes that are so equipped? That's right, very few pilots ever touch that control knob. So what if the ball is slightly off to one side of the trace? It's no big deal except that it does make for sloppy flying. I believe this is just one more example of the fact that so many pilots today ignore the rudder.

## IV The sight picture

In VFR pattern work, by stabilizing the airspeed with pitch (and adjusting as required with flap extension), the pilot is required to deal with only a single variable—altitude, or rate of descent. This the pilot controls with the throttle. Starting with the initial power reduction on downwind prior to any flap extension, as the pilot establishes a descent straight ahead, he or she continuously keeps asking, "Am I high, low, or just right?" Throughout, the pilot is glancing from airspeed to the runway, airspeed-runway, airspeed-runway. Of course, the pilot is also dividing attention, checking for traffic, etc. From the initial power reduction on, through the turn to base, final, and touchdown, the pilot makes whatever power adjustments are required to ensure that he or she runs out of airspeed and altitude simultaneously when the airplane is about 2 inches above the runway, and another "no-feel-the-touchdown" landing has occurred.

When the aircraft arrives at the point just above the runway, the pilot should break off the glide by bringing the nose up to the landing attitude (the "sight picture" the pilot established while on the ground) and holding it there. This is accomplished smoothly rather than abruptly to prevent a slight zoom, which would gain back several feet, at which time the airplane would abruptly quit flying and go "thunk." In order to hold the airplane in place just off the ground, the pilot must merely "take the slack out of the stick." By this I mean that as the airplane slows down, the yoke will tend to become loose in the pilot's hand. What the pilot must do is simply maintain a constant slight back pressure and attempt to hold the aircraft off the ground as long as possible. The result of using this technique is a "greaser" every single time. Try it. It works!

This "sight picture" technique works equally well, or perhaps even better, on the takeoff. As mentioned elsewhere, the modern approach to teaching takeoffs is to do it "by the numbers"—lining up on the runway, applying power, keeping an eye on the airspeed indicator, and when the magic number comes up, rotating. This system works perfectly for an air carrier or corporate jet, a light twin, or a high-performance single, but it's not the best way to do it in a simple training airplane. One student at a large flight school who was taught to take off by the numbers lost his airspeed indicator on the climb out and felt the need to declare an emergency! Needless to say, this caused a great deal of unnecessary activity at a very busy tower- controlled airport (the twenty-sixth busiest in the world). Had this student been taught that if the aircraft attitude is right, and the power setting is right, the airspeed must be in the ballpark, rather than having been taught to depend on his airspeed indicator to the exclusion of all else, he most certainly would not have felt the need to declare an emergency.

An entire class of aviation cadets, one of which was yours truly, during the Second World War went clear through their primary training (65 hours of flight, during which we learned to perform every maneuver for which the airplane was designed) without ever seeing the face of an airspeed indicator. For the purpose of this experiment, the instruments had been physically removed from the airplanes, leaving a hole in the panel. We did, in ground school, get to see a picture of one. The experiment, conducted by the United States Army Air Corp, was evidently quite successful, for at least the average number of cadets from this class went on to graduate and become military pilots.

How much simpler and more effective it is if the student pilot is taught the "sight picture" technique for takeoffs. Of course, flying by sight, sound, and feel won't work if one is required to get out of or into a short field. This kind of takeoff operation requires very precise airspeed control as the pilots climb out at Vx until all obstacles are cleared, then continue their climb at Vy until they have a goodly amount of space under them (the FAA recommends at least 1,000 feet).

In the final analysis, each pilot must decide for himself just what technique works best for him. He must do that which he finds is most comfortable for him, knowing that if he keeps the objective in mind and then does whatever is required to make the maneuver come out right at the end, he will have accomplished his purpose.

# V Sight, sound, and feel

Flying by "sight picture" as opposed to "flying by the numbers" really involves all the senses. Now we'll consider all the sensory cues pilots use in manipulating the controls of an airplane and making the machine do what they want it to do (or at least keeping it from doing what they don't want it to do).

The first of these is, of course, sight. Whether we are VFR or IFR, sight is the most important of the senses we use in flying. In VFR flying we keep the airplane upright by means of noting the position of the wings and nose of the airplane as they relate to the horizon. We not only control our pitch attitude by sight, but we also judge our angle of bank by sight. And when we're IFR, it is sight (the information we get from looking at the instruments) that enables us to remain upright. We simply must disregard the feeling we get through the inner ear and rely on what the gages tell us through our eyes. Also, the kinetic feeling we get from our deep muscles may be entirely wrong as a result of receiving false information from our inner ears. So much has been written about spatial disorientation and vertigo that we need not address that subject here. It is sufficient to note that a false sensation of turning when in fact we are flying straight can be almost overwhelming. It is at these times that we must force ourselves to rely on the messages from our instruments received through our eyes and interpreted by our brains (see Chapter 5).

We also gather useful information from the sense of sound. Any change in the engine sound as it drones along through the sky alerts us to a change in our flight condition. And if it gets really quiet in the cockpit, we know we're in deep trouble. If anybody should ask what a propeller on an airplane is for, tell 'em it's to keep the pilot cool, and if they don't believe you, just watch the pilot sweat when it stops! Adrenaline really starts flowing when a pilot hears the engine cough! When the engine skips a beat, it really gets our attention. We immediately start paying a lot more attention to the engine gages. Are the manifold pressure and tachometer needles steady? Are the oil pressure and temperature needles in the green? Any unusual sound and we are instantly alert.

The sound of the wind caused by our movement through the air also tells us things we should know. (Not nearly as well as it did in the days of the open cockpit airplanes.) There used to be a saying that as the tone of the music caused by the wind through the wires grew deeper, the pilot didn't need to worry until he started to hear "Nearer My God to Thee." Then it was time to bail out. Any change in the pitch attitude of the airplane will cause a change in the engine noise, and this provides useful data whether we are VF or IFR. Of course, pilots must process the data their senses send them for it to be useful. But even when we're IFR, a change in the engine sound may very well be the first clue we get that all is not well. At least it tells us that something has changed, and if nothing we did brought about the change, we'd better start looking for the cause.

Although pilots are admonished to not trust what they feel when they are in IMC (instrument meteorological conditions — in cloud), there are some kinds of feel (sense of touch) which are useful even then. For example, the tension (or lack of tension) one feels on the controls. Is the yoke getting loose in your hand, or does it seem to take more than usual pressure on your part to move the yoke? This kind of feel will alert us to a change in airspeed. In other words, pilots must force themselves (when in IMC) to ignore what they feel through the seat of their pants, but they should pay close attention to what they feel through the yoke in their hands. Is it becoming slack, or has an inadvertent increase in speed required us to exert more pressure to move it? Of course, you really can't tell anything if you have a death grip on the yoke. You must still strive for a light touch on all the controls, particularly the yoke and the rudder pedals. If the airplane is properly trimmed, a very light touch is all it takes to command the desired response. The old saying about holding the stick as if it was a little sparrow works in this situation. If you don't have a firm grip, it will fly away, but if you grip it too hard, you'll kill the poor little thing.

Now to expand on this business of sight, sound, and feel. With respect to sight, we've already had a lot to say about the "sight picture" by which we fly in the VFR environment. We have the whole world at our disposal, a 25,000-mile reference by which we can not only keep the airplane upright, but by which we can maneuver it and make it comply with our desires. This allows us to ignore the instrument panel except for the occasional scan to monitor the gages and see that all is well while we spend our time looking around outside, enjoying the scenery, and keeping the airplane on course by reference to ground objects, and, obviously, looking for traffic, which we can't avoid if we don't see. Of course, seeing alone is not enough. We have to understand what we're seeing, analyze and interpret the messages we're getting, and take appropriate action. And again, this is true whether we're VFR or IFR. Even when VFR, we use our sense of sight to monitor the gages that inform us of the state of health of the engine(s). If the heading or altitude should wander off, it is our sense of sight that tells us about it and tells us when we've made just the right adjustment to correct the situation.

Any change in the sound of the drone of the engine(s) will alert us to the fact that something is happening even sooner than the engine instruments themselves (manifold pressure gage and tachometer, but not the oil temperature, oil pressure, and head temp gages, or those instruments that advise us regarding the health of the electrical system). If the engine skips a beat, we hear it do so long before we see it on the tachometer (unless we happen to be watching the tach at the instant it happens). If there is a subtle change in the pitch attitude of the airplane, our ears alert us to the fact, and if we're in good VFR weather, we probably will notice the change in the sound of the engine before we catch it on the altimeter or the vertical speed indicator. So hearing can be quite important to a pilot. This sense may also be coupled with feeling to give us even more input. We "hear" by feeling a change in subtle vibrations.

Given enough experience, we all develop an "educated rear end." In other words we feel a slip or skid in the seat of the pants without the aid of the spirit level (the so-called ball-bank indicator or slip-and-skid indicator). Some people are more sensitive than others to this kind of unusual motion. For example, my wife, who has been exceptionally prone to motion sickness all her life, could always detect the slightest change in motion, but it took many thousands of hours of flying before I became as good at it as she is and always was. Now, however, if the airplane isn't perfectly balanced, if there's any slip or skid at all, I can feel it in the seat of my pants. I sometimes wonder at how some pilots can fly along in a steady slight skid or slip and be totally unaware of the condition. I've sat in airplanes and sworn that the pilot must be unconscious not to feel what's happening as the airplane gently slides sideways through the sky.

Another kind of sense of feel occurs when we encounter turbulence. After slowing to maneuvering speed and cinching the belt and harness up tight, we sometimes get a jolt that causes the head to meet the ceiling, and believe me, you feel that bump! A downdraft resulting in a sudden drop in altitude causes a sensation in the pit of the stomach similar to what is experienced in the start of a rapid descent in an elevator, or on the downside of a roller coaster ride. This, too, is feeling. Conversely, the added force of gravity ("G" load) felt with an abrupt pull-up or steep turn pushes us down in the seat and we definitely feel this, too. We know it when we get heavy. Here, too, the controls tighten up, just as they do with an increase in speed.

In addition to sight, sound, and feel, even the sense of smell enters into the equation. And although smell does not properly belong with sight, sound, and feel, still it should be covered here. How about that faint whiff of electrical fire? Or worse yet, gas or oil fire? I had a friend whose passengers got a strong odor of gasoline while flying at 7,000 feet, but he just kept going (he was only 30-odd miles from home and returning from a long trip). About then the engine in his 210 sputtered and quit. It died of fuel starvation, in spite of the fact that he was one of those guys who kept very careful records of the time in his tanks, rather than relying on notoriously unreliable fuel gages. He glided to an uneventful landing on a small, rough, sod private strip, which by happenstance was almost directly under him when the engine died. I happened to be overhead at the time and by chance I was monitoring the frequency, so I landed behind him and flew him and his two passengers out, but not before we found the belly of his airplane full of avgas. The fuel line that runs down inside the right door post had sprung a leak and all the gas (we drained out well in excess of a dozen gallons the next day) from his right wing tank had descended to the space between the cabin floor and the bottom skin of the airplane. Any errant spark could have spelled finis to the airplane and the occupants. Thank goodness none of the people aboard had decided to smoke on that flight! I later learned that the reason nobody aboard lit up was because of the strong odor of gasoline.

# VI  Power!

## The throttle

Ask most pilots to name the controls on an airplane, and the answer will be ailerons, rudder, and elevator. Some may mention the flaps, but very few will include one of the most important of all, the throttle.

An airplane's performance is based on the power it has available. Any increase in performance must be based on the difference between the power being used and the maximum power of the engine or engines, in other words, the remaining power available.

I'm certain everyone reading this has, at one time or another during his or her training, heard the expression "being behind the power curve." But how many of you really understand this concept? Let's have a show of hands. Okay, those of you with your hands up may be excused from reading this chapter, both of you. All the rest, those who didn't raise their hands, stick around and pay attention. You might learn something. Once again we are discussing that difficult-to-visualize concept—angle of attack.

First off, a better way of expressing what happens when you are behind the power curve is covered by saying you are in the "area of reverse command." By this we mean in order to make an airplane climb, you must lower the nose, and if that isn't reverse command, I don't know what is! Any further decrease in airspeed will require the addition of power, another example of reverse command! You get into this situation by increasing the pitch with that poor little engine putting out all the poop it can muster until the angle of attack is just slightly below the stalling angle. At this point, the airplane will no longer climb, but will just sort of hang there. (The expression is "hanging on the prop.") You are now in the area of reverse command, and if you want to go up you will have to lower the nose, and if you want to go slower, you'll have to add power. Both cases are just the opposite of what you normally do when initiating a climb, or starting to slow down, hence the term "area of reverse command."

In this flight condition with an extreme nose-up pitch and full power, P-factor (including torque and corkscrew effect of the slipstream from the prop) is at the maximum, and you'd better be right on top of your ability to use the rudder, otherwise if the airplane should stall with the built-in yaw and full power, it will fall into a spin so fast you won't know what hit you.

Actually, there's only one reason a pilot would ever want to put himself behind the power curve, or in the area of reverse command. At least I can't think of any other valid reason, and that's when making an extremely short landing. But I do believe it is important for every pilot to know and understand this phenomenon, what causes it, and what has to be done to get out of it. (Reduce the pitch and let the airplane fly out of this situation.)

A good number of airplanes have come to grief when the pilot attempted to climb out of a short field and over an obstacle by increasing the nose-up pitch too much rather than maintaining the published Vx (best angle of climb speed). When the pilot discovers that he isn't climbing steeply enough to get over the wires, trees, or whatever at the end of the runway if he or she increases the pitch until the area of reverse command is reached, the airplane may continue to go, but what it won't do is go up. Remember, whatever performance the airplane has in terms of gaining the most altitude for a given amount of forward distance is at Vx. Go faster and you won't get as high. Go slower and you won't get as high as you cover a given distance.

In a simple (fixed-pitch, fixed-gear) airplane, power management is fairly simple. Generally, one uses full power for takeoff and climb and another for cruise, something less for descents, and then a gradual power reduction for landing approach. However, the minute you step up to an airplane with an adjustable (constant speed) prop, you are confronted with an entire range of power settings. Maximum for takeoff, something less than that for initial climb, and perhaps still less for cruise-climb, and a great many choices for the cruise itself. For most efficient cruise, as a rule of thumb, it is best to select the lowest rpm, coupled with the highest manifold pressure permissible. Did you get that "permissible" in there? This simply means that if the manufacturer says its okay to exceed the rpm in hundreds with the MP in inches (over squaring the numbers), you may do so.

## Mixture

While we're on the subject of power we might as well mention a related subject that is frequently overlooked by many flight instructors. We're talking about leaning. I've had several pilots tell me that during primary training they were told to lean until the engine starts to run rough, and then "rich it up a little past the point where it smooths out." But when they transitioned into an airplane so equipped, they were left on their own to figure out how to use an EGT (exhaust gas temperature) gage, which is a much more scientific, precise, and accurate means of leaning the mixture. Like the true airspeed indicator, the EGT is something that most pilots have to learn themselves, but at least in this case there are printed instructions. And, of course, leaning becomes extremely important when taking off from a high-altitude airport when you want all the power you can get. And contrary to what most pilots are taught during their primary training, you should lean from the ground up anytime you are operating at or below 75 percent power, not just when you're above 5,000 feet. (When the fuel crunch hit everybody in 1972, especially those of us in general aviation, Avco Lycoming (manufacturer of engines for light aircraft) sent Joe Diblin, an engineer and editor of their house publication on engine care, all around the country preaching to pilots the fact that they should lean from the ground up anytime they are operating at or below 75 percent power, and to disregard what they had been taught— to only lean above 5,000-feet pressure altitude.)

The best reason I can think of as to why instructors teach their students to lean only when they are operating at and above 5,000 feet is that they are insulting the intelligence of their students, believing that the students are too dumb to remember to rich up the mixture when descending. Why they are expected to remember when descending from 5,000 or above, but not from, say 3,000 is beyond me.

## VII  Speed

Every one of the Practical Test Standards for both airplanes and gliders has a TASK requiring the applicant to demonstrate flight at "minimum controllable airspeed." Most applicants manage to pass this part of the test, but then they promptly disregard whatever they have learned about controlling the aircraft in the slower-speed modes.

I recently spent a week flying with a guy in his airplane, a gentleman whom I had certified for his instrument rating about 20 years ago. Henry has owned and flown a super-duper, extremely well equipped Bonanza for several years, even flying it to Europe and back and to South America and back. Obviously he is quite a competent pilot, but so help me, his landings are awful! He crosses the fence at 120 (when he should be slowing down from a final approach speed of perhaps 80) and plasters the airplane on the runway with 10 degrees of flap! This frequently results in a series of skip-type bounces, and as I watched this repeatedly not once did I see him add a touch of power which would have put a cushion under him as he eased back to the ground from a moderate bounce. Try as I might, I have so far been unable to get Henry to slow the airplane down. He says he was taught to use only 10 degrees of flap for landing because he will get better climb performance if he has to make a go-around. I tried to tell him that if he goes to full flaps once he is firmly committed to land, his landings will improve, but I couldn't get him to do it even after I demonstrated a few landings from slower approach speeds and full flaps, a greaser every time. I tried to explain that the FAA says a normal landing includes the use of full flaps, but to no avail. With some of the hard bounces I watched him do, it's a wonder the landing gear struts didn't come up through the wings. Beech just builds a sturdy airplane!

Several years ago the FAA conducted a series of tests designed to measure the effect of excess speed, and it was determined that a mere 5 knots too fast over the threshold is the equivalent of being 50 feet high, and this translates to the need for a much longer runway. So you can see how important it is to be comfortable in the slower speed ranges.

As a flight instructor, whenever I administer a flight review, I have the reviewee demonstrate flight at minimum controllable airspeed, just hanging on the edge of a stall, and after he or she has stabilized the airplane in this condition straight ahead, I ask him or her to make a few 90-degree turns each way, usually with the stall warning device (either a horn beep—peeping away, or the light flickering on and off) activated. When a pilot becomes comfortable in this situation, he or she is reaching for the outer limits of the airplane's performance envelope and can safely operate a bit inside this extreme.

I realize that the controls are looser, and the response is nowhere near as crisp as it is at higher speeds, but when a pilot is confident of his or her skill in this area, he or she will be an infinitely better pilot.

For a guy who advocates using the "sight picture" and sense of feel in landing training for primary students flying fixed-pitch, fixed-gear airplanes, it would seem as though I would not be a strong advocate of precise airspeed control in other areas of flight, but I am. It is absolutely essential that pilots be precise in their airspeeds when flying a complex single, and even more so in a twin.

# 3

# Stalls, spins, and other bad stuff

## I Stalls

Flight schools and flight instructors are doing it all wrong! What is being taught is how to make an airplane stall and then how to recover. What should be taught is stall recognition. But before even that is possible, a thorough understanding of the stall is required. And this means we must go back to basics and review the concept of angle of attack, what it is, what creates it, and how it is affected by pitch and power.

Since we can't see the two lines that intersect to form the angle, this concept of angle of attack seems to be particularly difficult to grasp, probably because it deals with air, which is invisible. Reduced to its simplest, and disregarding angle of incidence, the angle of attack may be viewed as the angular difference between where an airplane is pointed and where it is going. An angle, any angle, is formed by the intersection of two lines. The two lines that form the angle of attack where they meet are the relative wind and the MAC. (No, I don't mean a great big hamburger; I'm referring to the mean aerodynamic chord of the wing). The relative wind is parallel but opposite in direction to the flight path of the airplane. In other words, where the airplane is going, not where it's pointing.

Now, for each airplane there is a specific angle of attack at which it can no longer sustain flight. This is a fixed number and it never changes. (In most lightplanes, it is somewhere in the neighborhood of 17 degrees.) Angle of attack is built into the design of the airplane. Below this critical angle of attack, the airplane cannot stall. Exceed this magic number, and it will stall! It is at this point that the airflow over the wing separates from the wing's surface and it begins to burble. Every student pilot is taught and understands this part, but if you ask a pilot what makes an airplane stall, the odds are you'll get an answer based on lack of airspeed or air burbling over the wing, or the pitch attitude got too high, all of which may be true, but all of which fail to answer the real question. An airplane stalls because it has exceeded its critical angle of attack. That's simply all she wrote! An airplane can stall at any airspeed and at any attitude and any power setting!

If pilots would quit thinking in terms of air flowing over the wing and gear their thinking to considering the air as striking the underside of the wing, they would be better able to grasp the concept of angle of attack. Perhaps the difficulty lies in the fact that we can't see the two lines which converge to form the angle of attack. This is probably what causes the conceptual problem in the minds of most pilots. Where most aviation educators (including many knowledgeable aviation writers, and even including the FAA in some of its publications) go astray is in the fact that they relate the stall to airspeed. Since most general-aviation lightplanes don't have angle-of-attack indicators, for want of anything better, stalls are related to airspeed, but if you think about it, the trigger that sets off the stall warning device, be it aural or visual, is really an angle-of-attack indicator. Corporate jets, military aircraft, and air carrier airplanes are equipped with angle-of-attack indicators, but even though they are cheap and easy to install, general aviation singles and light twins have to struggle along without them.

With all the literature emphasizing "stall speeds," it becomes very difficult to impress on the student pilot (and even the experienced pilot) that the published stalling speeds are only valid under a very narrowly prescribed set of circumstances. All too often pilots marry themselves to the idea that an airplane will only stall if it gets below a specific, published airspeed, and if they just keep the airplane above that magic number it won't stall. Nonsense! It can stall at any airspeed.

This is why we used to teach the "accelerated maneuver stall"—to impress on the student that the airplane will stall at something other than the published "stalling speed" and to demonstrate the effect of load factor on the stall. Unfortunately, the accelerated maneuver stall has been deleted from the Private and the Commercial PTS (Practical Test Standards) and thus from the respective curricula. The thing to remember about this maneuver is that although the airplane's attitude changes, the all-important flight path does not.

The country's leading aviation writer has stated that stall recovery involves "lowering the nose to reduce the angle of attack and increasing power to gain speed." This is totally false! It is true that lowering the nose of the airplane will reduce the angle of attack, but so will the addition of power. The relationship between power and angle of attack is rarely taught, and it certainly should be. Every publication I've ever seen and most instructors with whom I've discussed stalls teach stall recovery by exhorting the student to reduce the angle of attack by lowering the nose, and then adding power to increase airspeed or to conserve altitude, or prevent a substantial loss of altitude. This is not why we add power in a stall recovery. We add power to reduce the angle of attack. Remember, the angle of attack is the degree of difference between where we're pointing and where we're going, and when we add power, we change the direction of travel and thus we change the direction of the relative wind, reducing the angle of attack.

Anytime you add power, if you don't change the pitch attitude of the airplane, you are reducing the angle of attack. Think of it this way: An airplane in a level attitude with reduced power is pointing straight ahead but is going forward and down at some glide angle. Now, to stop the descent, what do you do? You add power and drive the airplane straight ahead. What you have also done is to zero out the angle of attack because now you are going right where you are pointing. Another graphic demonstration of the relationship between power and angle of attack can be seen in the difference in pitch-up (deck angle) between a power-on stall and a power-off stall. The airplane stalls at exactly the same angle of attack, but that angle is reached with a much greater pitch-up with power than without power. If we hold the pitch attitude constant and add power, we have reduced the angle of attack without doing anything else. What we've done is change the flight path of the airplane, thus changing the direction of the relative wind. And what has this done? It has reduced the angle of attack!

Anytime you make a power adjustment and do not change the pitch attitude, you are changing the angle of attack. Since we can't change the chord of the wing, what we are doing is changing the direction of the relative wind by changing the flight path of the airplane. You can prove this by entering a power-off stall and holding the yoke exactly where it is as you add full power. The airplane will recover from the stall without lowering the nose! And this whole concept of relating stalls to angle of attack without regard to airspeed is not adequately emphasized by most flight instructors.

Over the years I've heard many fancy names applied to stalls which occur under differing conditions; landing attitude stall, takeoff and departure stall, approach to landing stall, and accelerated stall, to name a few. This has tended to lead our students away from the fundamental fact that, as Gertrude Stein might put it, a stall is a stall is a stall is a stall, and anytime the critical angle of attack is exceeded, the airplane will stall, and if that critical angle is not exceeded, the airplane simply cannot stall!

Of course, there is real value in teaching the so-called approach to landing, takeoff and departure stalls, for it is in these phases of flight that a stall can be most dangerous—when you're low and close to the ground and may not have enough altitude under you for recovery. Both of these stalls should be practiced in turning flight, which also loads the airplane with excess weight. Plus, in curved flight you're never going right where you're pointing.

In the approach to landing stall, at a safe altitude, the airplane is configured for landing, and then the turn from base leg to final is simulated, with a very tight turn to simulate overshooting the runway-extended centerline. This is a classic example of a dangerous situation. The takeoff and departure stall should also be practiced in turning flight. With full takeoff power, again at a safe altitude, that first left turn in traffic should be simulated as well as a distraction causing the pilot to divert his or her attention as he or she continues to apply elevator pressure. In both of the above cases recovery should be initiated at the first positive indication of a stall, and this doesn't mean when you see or hear the stall warning device activate. Rather, recovery should be initiated at the first sign of the prestall buffet. Recovery, of course, consists of simultaneously reducing the up elevator pressure, leveling the wings, and adding all available power. If the power is already all there, you can't add any more, but if there's anything left, get it in!

Although instructors try valiantly to impress upon their students the fact that the published stalling speed of a given airplane make and model is a variable, they are often unsuccessful. Somehow students (and experienced pilots as well) get it in their heads that if they only keep the airspeed above that magic number, the airplane won't stall. Wrong! Of course it can, and it may if other factors enter into the equation. This is why accelerated maneuver stalls should be taught — to demonstrate that if the load factor is increased, the stall speed goes up. And the markings on the airspeed indicator are no help either. The bottom of the green arc and the bottom of the white arc are indicative of a stall only under a narrowly prescribed set of circumstances. For instance, in each case the wings must be level.

In a steady, unaccelerated flight condition, the published stall speed will be close to accurate, and this applies to indicated airspeed, regardless of what the true airspeed actually is. An airplane operating at a high-density altitude will stall at a substantially higher true airspeed than the same airplane being operated at sea level. This is easily demonstrated by stalling an airplane at a low, safe, altitude, noting the indicated airspeed at the break, then climbing several thousands of feet and repeating the process. The indicated speed at which the break occurs will be identical, but the true airspeed will be much higher up there in the thinner air. It's like taking off at a high-elevation airport. The airplane breaks ground at a much higher actual speed than it does at a lowland field, but the airspeed indicator will read exactly the same. Thus it can be seen that the airplane reacts to indicated airspeed. That's all the airplane knows. And that's why so much longer runways are required at high-elevation airports — to allow room for the takeoff and for the landing roll. (Since you are landing at a much higher actual airspeed than what is indicated, you require a good deal more distance for the landing roll.)

Buz Massingale has an interesting technique for teaching density altitude. He compares it to the molecules of air in a phone booth. The airplane gets its lift from the number of molecules of air in a cubic phone booth, and if the phone booth is occupied, the body heat of the occupant will drive out some of these valuable molecules. And as the occupant begins to sweat, the air in the phone booth will hold more moisture. Since water droplets, unlike air molecules, don't provide lift, the molecules of air displaced by the moisture will leave less molecules per cubic phone booth, and thus less lift. How's that for simplification?

The method by which students used to learn that an airplane can and will stall at something other than the published "stalling speed" was by means of demonstrating and teaching the accelerated maneuver stall. However, in its infinite wisdom (stupidity?) the FAA has deleted this maneuver from the requirements of the Private Pilot PTS

(Practical Test Standards) as well as from the Commercial Pilot PTS. Thus, it is no longer taught. Russ Glover pleads:

> "... for the re-introduction of the accelerated stall into (the) PTS. I believe it is possible to become an airline captain without demonstrating accelerated stalls of any kind, or spins, or any meaningful unusual attitudes. Airlines of late have been rolling inverted and diving straight in for a variety of reasons, but the passengers have a right to expect that the captain has been inverted a few times in his life and might have a fighting chance to roll the critter right side up. (I) suggest a five hour basic aerobatic course (as a) requirement for the ATP! If the poor guy barfs or can't do a simple recovery from sustained inverted flight in something like a Decathlon or T-34, he ain't qualified to fly passengers for hire."

Another valuable lesson dealing with stalls that is rarely taught anymore is the so-called delayed recovery stall, where the student gets to experience a secondary stall, and no doubt the deepest stall he will ever see. This maneuver also impresses on the student the value of good rudder usage.

It is accomplished like this: After climbing to a good safe altitude and after first carefully clearing the area where you are working, apply carb heat (if called for in the airplane you're using) and retard the throttle to idle. Now slowly and smoothly bring the yoke all the way back to the stop, being sure to keep the wings level and the nose pointed straight ahead with the rudder.

The nose will pitch up and the airplane will stall. Hold the yoke full back and watch what happens. Remember, you must keep the wings level and guide the airplane straight with the rudder. If a wing starts to fall off, catch it immediately with opposite rudder. When the stall breaks, do not relax the back pressure. Keep holding the yoke full back, right to the stop. The nose will fall through the horizon, and the airplane will recover by itself. It will then pitch up and stall again and repeat the process through a series of what are called *phugoid oscillations*. If you maintain heading and wings level with the rudder, it will keep this up all the way to the ground.

This delayed recovery or secondary stall training is one of the greatest confidence-building maneuvers I know of, and it really makes an impression on the student. When a student has progressed to the point where he or she can experience this, he or she will become convinced that airplanes are indeed safe. If you are fortunate enough to be among those who have had this sort of "delayed recovery" stall demonstrated, you are a member of a very small minority and are fortunate indeed. To further explain what is going on here, you should know that this phenomenon is a result of the fact that the tail (horizontal stabilizer or stabilator) is in and out of the stalled condition, as well as the wing.

## II  Should we reintroduce spin training in the primary flight curriculum?

The answer to this question is maybe yes, maybe no. It all depends on just how spins are taught. It would be a wasted effort to have every pilot learn to spin an airplane and recover if the technique is taught the way most instructors today learn spins. It goes like this: The regulations require that an instructor applicant either demonstrate spins on his or her CFI flight test or produce a logbook entry stating that the applicant "has had instruction in spin entries and recoveries left and right." So, sometime during training for the CFI, the student's instructor says, "Today we're going to do the required spin work so I can make an appropriate endorsement in your logbook."

They then get into the airplane and take off, climb to a nice, safe altitude and the instructor advises the instructor student to watch closely as he demonstrates the required spin entries and recoveries. The two elements required for an airplane to spin are: one—it must be stalled; two—there must be a yaw moment introduced. So what the instructor does is apply carb heat (as required for the airplane being used), retard the power, pull the yoke back to get a sharp pitch up, and just as the stall break occurs, the instructor mashes on the left rudder, holding the yoke full back. The airplane responds by falling off to the left and starting to rotate. After about a quarter to one-half of a rotation, the instructor applies hard right rudder. He follows this by relaxing the back pressure on the yoke. The airplane responds to these control inputs by stopping the rotation and becoming unstalled. But since it has been in a stalled condition, with the nose down, it is now in a steep power-off glide, almost what could be called a dive. So the instructor then comes back on the yoke to recover from the dive, and as the nose comes up, the instructor applies power. The instructor then levels off and turns to the instructor student and says, "There, that was a spin entry and recovery to the left. Now we'll do one to the right."

The instructor then repeats the process, only this time applying right rudder to enter and left rudder to recover. And when they get back on the ground and return to the flight school office, the instructor makes the appropriate entry in the applicant's logbook, stating that this individual has had instruction in spin entries and recoveries both left and right.

But has the applicant really had instruction? What did the applicant learn? Probably that the instructor was scared of spinning an airplane and so the applicant should probably be scared too. What good will this kind of "instruction" do if the applicant should blunder into an inadvertent spin with a student of his or her own? None, that's what! I suppose this technically satisfies the requirement, but with this kind of so-called spin instruction, it is doubtful if the new instructor would even be able to recognize an inadvertent spin, let alone recover. Unfortunately, however, this is what too often passes for spin instruction for the instructor applicant. That's why I say that unless it is accomplished in a completely different fashion, spin training is a total waste of time, effort, and money. And it may even be detrimental.

A few years ago I knew a young flight instructor who got his spin training in just that way, and he hated spins. Said he'd never purposely do a spin again. In fact, when he trained an instructor applicant, he turned the applicant over to another instructor for the spin training and logbook endorsement. In the first six months of his instructing career, this young instructor who hated spins found himself in no less than four inadvertent spins caused by students mishandling the controls during stall training, almost certainly because of improper rudder usage. I have been instructing for over 50 years and it only happened to me once, and that was because I let a student go too far before taking control. Inadvertent spins are invariably the result of improper rudder usage. What I have described above is certainly no way to teach spins.

On the other hand, if spin training were to be administered properly, I would strongly favor putting spin training back in the primary flight curriculum. By done properly I mean to teach spins and recovery from normally anticipated flight situations. Here are a couple of scenarios that illustrate how it could be done:

One, at a good safe altitude, after carefully clearing the area, the instructor should have the student put the airplane in a steep bank, say 50 or so degrees of bank. In this condition, to prevent the overbanking tendency which wants to keep the airplane rolling, a certain amount of aileron must be held against the bank, or toward the outside of the turn. Now to compensate for this, a certain amount of bottom rudder must be held, toward the inside of the turn, thus setting up a "cross control" situation. This is okay as long as we maintain coordination and keep the ball in the center of the trace in the ball-bank (turn and bank, slip and skid, turn indicator, or whatever the FAA is calling it this week) indicator.

However, if the steep turn is uncoordinated, we are inviting trouble. In addition to this we have increased the power to offset the loss of the vertical component of lift, which, because of the bank, has gone off on a diagonal. This results in the addition of P factor and torque. Now just a bit of back nudge on the yoke or stick will result in an accelerated maneuver stall because we have drastically increased the load factor by banking so steeply. Since we already have a built-in yaw as a result of holding the rudder, plus the added P-factor we have the two factors required for the spin, a stall coupled with a yaw. And since we are doing this with a lot of excess power, the airplane will respond by briskly rolling toward the inside of the turn, and it will commence to spin at a rapid rate. The first time this is experienced, it is guaranteed to startle the pilot, to say the least. This is called an "out of the bottom" entry. If top, or outside, rudder is held, the low wing comes up and over before the airplane commences to rotate, and we have an "over-the-top" spin.

The recovery from either of these spin entries is simple and quite easy. If the student is well briefed beforehand, the instructor can talk him through the entire procedure without even having to demonstrate, but since the entry is quite startling, it is better if the instructor demonstrates one first before having the student actually do it. The first step in recovering from such a spin entry is to instantly kill the power. I mean cut it right back to idle! This is followed immediately with the hard application of full rudder against the direction in which the airplane is spinning. This stops the rotation, and recovery from the ensuing stall is routine. Because of the speed at which such an entry takes place, the student will get to experience at least a two-turn spin before recovery is effected.

This kind of spin entry should be taught from both left and right turns so the student can graphically see the difference in response between the left and the right entries. In the left, P-factor and torque are at work aggravating the situation, while in the right entry this element is working against the tendency of the airplane to spin, and some little effort is required to force the spin entry.

Another scenario—and one that is based on a situation that is much more common for unwary pilots to find themselves in by inadvertence—is setting up, again at a nice, safe altitude with the area well cleared, a tight base-to-final turn. If this is started at 5,500 feet, ground level is hypothesized at 5,000. At this point a glide is established straight ahead. Then a left turn is initiated, simulating the turn from base to final. Since the turn to final is hypothesized as overshooting the runway, the turn is tightened up with a steeper bank and more back pressure on the yoke. The back pressure is kept up until the airplane stalls. If bottom, or inside rudder is held in the turn, a spin "out of the bottom" will result. This kind of spin entry is quite realistic and is responsible for most of the fatal stall-spin accidents.

The recovery is quite standard, while not really important because if a spin is entered in the pattern as this one simulates, there would be no space for recovering. If flaps have been applied, they should be retracted at once as full, hard opposite rudder is briskly applied right to its fullest extent. When the rotation stops, standard stall recovery technique is used. The flaps, if extended at the onset, must be retracted, for the speed build-up in the recovery dive could well exceed the flaps-extended speed limit.

Still another very realistic situation can be set up at altitude, again hypothesizing a landing approach, but this time on short final. With the airplane trimmed for landing, just as the mythical runway is approached (some 5,000 feet above ground level), a go-around is initiated.

With the application of full power and the airplane trimmed nose up for landing, a bit too much back pressure will cause the airplane to stall, and if right rudder has not been applied to counteract the P factor, the sudden application of full power will cause a yaw to the left. This coupled with the stall has introduced the final element required for the airplane to start to spin, and since we now have climb power, it will do so in a hurry. Again, the quickness with which this happens is guaranteed to startle the student.

Once again, recovery is fairly normal. With the instant reduction of power, the flaps, if extended, should be simultaneously retracted, followed by the hard, full rudder deflection against the direction of turn, followed by a normal stall recovery. However, in this situation again, recovery, except in practice, is not important, for in a real situation like this there would be no space for recovery and death would be the likely result!

These are just a few of what I mean by spin training in entries and recoveries from normally anticipated flight situations. This kind of spin training can be very beneficial for any pilot. However, if the fact that an airplane cannot spin if it is not stalled is kept in mind, and the pilot is well trained in stall recognition as opposed to stalling and recovering, the pilot is not likely to be confronted with any situation where spin recovery is required. What we have really taught is stall, and consequently spin, avoidance.

For those of you who are into the study of aerodynamics, let me point out that I have ignored the actual mechanics of the spin phenomenon such as the fact that one wing is in a deeper stall than the other, in order to concentrate on what the pilot needs to know about spin entry and recovery. While adequate to educate the pilot regarding spins, a good ground session is helpful, but it is no substitute for actual training and practice in the air. Please understand, none of these scenarios should be attempted without a knowledgeable and experienced instructor aboard to demonstrate them. They are not intended for student pilots to attempt solo based only on the descriptions found here. Most inadvertent spins result from improper rudder usage, and most inadvertent spins occur too low and close to the ground to allow sufficient space for recovery. All general-aviation lightplanes will recover from a spin by themselves if the pilot can force himself/herself to just let go of everything, all the controls. But this technique of spin recovery requires more altitude than does the positive input of control force. And it requires nerves of steel and patience on the part of the pilot.

Done properly, spin training is a desirable addition to the training curriculum. Not done properly, it is not only a waste of training time, but may very well be an actual detriment to the making of a pilot. While the goal should be spin avoidance, this can be best accomplished if the pilot is well trained in stall recognition and spin entry and recovery, and especially in rudder usage.

# III The steep spiral

At least as dangerous as the spin, and equally as easy to blunder into by inadvertence, is the steep power spiral. In fact, some airplanes which by design are somewhat spin resistant wind up in a spiral when the pilot is attempting to make them spin. A steep spiral may appear somewhat similar to the spin, but there is a big difference. In a spin, the airplane is stalled, while in the spiral the wings are flying rather than stalled.

What makes this maneuver so dangerous is the fact that enormous loads are placed on the airplane when the pilot attempts to abruptly pull up, up to and including the point at which the metal yields and comes unglued. This is what happens: In a medium banked turn, if the nose of the airplane is not held up with elevator pressure, it will tend to drop, causing the speed to increase. This, in turn will cause the bank to steepen due to the fact that the higher speed of the outside wing has given it a greater angle of attack and thus more lift than the inside wing. Now the effect of cumulative causation comes into play. The greater bank causes the pitch to increase, which in turn causes the bank to steepen further, which causes a still greater pitch increase and so on, *ad infinitum*.

And there's still another factor at work here. The spiral is aggravated by the fact that an airplane is pitch stable, both in level flight and in a spiral dive. Our friend the trim tab attempts to return the airplane back to the trimmed airspeed, and in doing so tightens the spiral.

Sometime during this process, the pilot decides to stop this undesirable state of affairs, so what does the pilot do? He or she hauls back on the yoke (or stick) in an effort to bring the nose of the airplane back up, and with the extreme added load factor, the pilot is likely to pull the airplane apart! This is particularly dangerous when the pilot has no outside visual references (see the section on Priorities below).

To avoid this, it is absolutely essential that the first step be to reduce the power all the way to idle. Then the bank should be eased off until the wings are level. Then, and only then, should an attempt be made to bring the pitch back up to level, and again this action must be gentle and easy. No abrupt or forceful actions allowed! Finally, after the wings are level and the nose is back up to normal, power can be reapplied, smoothly. And there's another undesirable result of the pull-up from a steep power spiral. Even if there is no apparent structural damage, there may very well be some degree of hidden stress damage which can show up later (perhaps with another pilot flying the airplane) at an equally bad, or even worse, time. This kind of hidden stress damage tends to be cumulative, with each episode weakening the structure further until just a bit more causes something bad to happen—a necessary component, such as a wing or part of the tail, breaks off the airplane. And we all know they don't fly very well without these parts being firmly attached to the rest of the structure. This cumulative buildup toward total stress is the same thing that happens when we bend a piece of metal or wire back and forth until it breaks.

# IV  Priority
## Disorientation in cloud

One situation in which it is absolutely essential that a pilot perform an operation "by the numbers" occurs when a nonqualified pilot blunders into IMC (instrument meteorological conditions). The scenario goes like this: When the pilot loses visual contact with everything except perhaps his or her own wingtips, just the weight of the hand on the yoke pulls it down slightly. This causes the airplane to bank a bit to the left, resulting in a slight pitch down. This results in an increase in the bank, leading to a further increase in the pitch down, and so on, until the airplane is in a steep grinding spiral power dive toward the ground, a so-called "graveyard spiral." By actual test during the Second World War, using 10,000 aviation cadets as subjects (I was one of 'em!), it was discovered that this process takes a grand total of 22 seconds for a fully developed graveyard spiral to occur.

About this time the pilot glances at his airspeed indicator and says to himself, "My God! Look how fast I'm going!" He then grabs the yoke with both hands and gives a good, healthy tug. When they find the wreckage with the bodies, I'm talking about dead bodies here, the airplane is scattered over the countryside in pieces. And the newspapers report another in-flight breakup. And another airplane manufacturer gets unjustly sued for the "faulty design" of the airplane.

What the pilot must do in this situation, in sequence, is to instantly kill the power. I mean all of it, right back to idle. Next, the wings must be leveled up, easily and smoothly. Finally, the nose must be eased up, and I do mean eased. No jerks or tugs allowed, just an easy slow pull-up.

This is the classic "unusual attitude" in which pilots find themselves, and the only possible recovery is as described above, first power, then bank, and finally pitch. Although most flight instructors teach this, it takes real nerve to steel oneself to execute it properly in the real situation, but it is imperative that it be done in just this way if survival is to result.

If, however, the airspeed is low and getting lower, the first step is to reduce the pitch angle (lower the nose), get the wings level, and come in with enough power to bring the nose to level and keep it there. By following these simple steps in the proper sequence, recovery will result from the two classic unusual attitudes pilots are likely to find themselves in, and control will be regained.

# V Crash survivability

No one ever takes off expecting to crash, but like fire drills, if pilots prepare intelligently for the worst, they can maximize their survival potential. Even if pilots do everything right in the proper sequence, sometimes pilots find themselves faced with the prospect of ending up on some unsuitable piece of geography. The loudest noise in aviation is the sound of silence when the engine quits unexpectedly in flight, leaving the pilot with more questions than answers, and a suddenly very busy agenda. As pointed out in Chapter 16, the prospect of safely making the subsequent landing must be the primary concern for the pilot, whether or not he or she is successful in effecting a restart. Of course, the most important thing the pilot can do at any point until landing the airplane is to keep his or her cool and fly the airplane. Do not under any circumstances permit the airplane to fly you!

If a restart cannot be accomplished and a suitable landing area is not available, there are several things the pilot must do to get down to dear old mother earth with a minimum of damage and personal injury. This will not only save damage to the human body, but it will also make your insurance carrier happy, or at least happier than it otherwise would be, an outcome certainly to be desired.

Two distinct elements determine crash survivability: deceleration forces and maintenance of occupiable space. First, of course, the pilot must continue to fly the airplane all the way to touchdown. The big killer in a landing accident isn't the airplane colliding with the ground, but the resultant collision of the occupants with the aircraft interior or loose items in the aircraft. Avoiding encounters with lethal objects requires that proper restraints be used for all personnel aboard and all stowed objects.

Seats, seat belts, and shoulder harnesses work well in forward-impact situations but are nearly useless in high vertical impact cases. Hence, it is imperative that the airplane not be stalled prior to impact. The sudden stop after coming down flat after a stall is likely to result in spinal compression and death!

Forces which cause massive deformation (collapse) of the cabin reduce the occupiable space to unsurvivable proportions. An examination of agricultural aircraft discloses that the pilot compartment is built with a cagelike structure designed to retain its shape under impact, but many general-aviation aircraft are poorly designed for sustaining impact forces around the cabin area. When a forced landing on unsuitable terrain is required and the pilot must decide what to hit, the object should be to avoid hitting anything with the main part of the fuselage. The landing gear and wings should be made to take the brunt of the impact forces, and if possible, to shear off, thus preserving occupiable cabin space. I'm sure you are familiar with the old story about the student who was told to be sure to put the airplane between two trees so the wings can take the impact in case you have a forced landing. One day when the student pilot was out practicing solo, his instructor got a call to come and get him. He had just crashed and was uninjured because he had followed instructions and put the airplane between two trees and the wings had broken off. When the instructor arrived at the site of the crash, he found the airplane in a huge, flat field, in the center of which were the only two trees around, and by which the wreckage was located!

To work properly, belts and harnesses should be tightened so as to hold the body motionless at and after impact. This means they should be uncomfortably tight. When you encounter turbulence and the airplane starts to get bounced around, you cinch up the belts, don't you? Well, do the same thing when you know you're going to hit something hard (and this includes water). Make sure that all baggage and loose objects are well secured, as they can become dangerous missiles during the crash. Stowing headsets can also eliminate another source of injury.

A forced landing is never a comforting prospect, but even under the most hostile of circumstances, it can be survivable if executed properly. By intelligently using the laws of physics, a pilot can maximize his chances of walking away from the aircraft. The key is how the energy is dissipated in the fast-moving vehicle to bring it to a stop. As any skydiver will wryly tell you, it's not the fall that gets you; it's the sudden stop at the end. The same is true of any aircraft landing. Ideally, the engine quits in a position where a squeaker landing can be made on a runway and a clean turn-off can be made before rolling to a stop. (I did that once when the airplane I was flying blew a jug on downwind.) But more often the choice of landing sites isn't that good. At the other end of the spectrum, an aircraft colliding straight ahead into a sheer mountain face is pretty much guaranteed to result in the fatality of all occupants.

The aircraft has a lot of stored-up energy as a function of its motion through the air, and according to the laws of physics, force equals mass times acceleration. For a given mass (loaded airplane), the faster it goes, the more force is stored up. In bringing all that to a stop, a whole bunch of force is released. The key to keeping the deceleration forces minimal is to avoid any sudden stop. Using a flat surface with no obstructions and landing with a shallow approach angle will mean a longer rollout and more gradual dissipation of impact forces. If there are obstructions, it helps if they are frangible (breakable), such as corn stalks, small trees, or saplings which dissipate energy as they break. The same is true concerning frangible parts of the aircraft itself, like wing spars, landing gear, or other struts.

High-angle impacts are to be avoided at all costs, as there is less distance available to bring the aircraft to a stop. This results in higher deceleration forces. This type of impact occurs when pilots attempt to slow the airplane too much and stall. The airplane should be slowed down as much as possible, so that it is moving as slowly as it possibly can at the moment of impact, but this does not mean it should be allowed to stall. If the aircraft is permitted to stall, the resultant pancake landing, even from a low altitude, nearly always results in serious injury or fatality from spinal compression.

Bear in mind that the best glide speed provides the least sink rate. The glide ratio (forward distance covered per unit of descent) is the reciprocal of the impact force in G (unit of gravity). Thus, a glide ratio of three to one yields an impact force of three-tenths of one G. Not too hard on the human body, huh? The object, then, is to be going forward and down as slowly as possible when you encounter some unfriendly object, such as rocks, trees, brush, or whatever.

## VI How much can they stand?

During the Second World War, I saw airplanes come home from combat missions with absolutely unbelievable battle damage. Of course, military aircraft are built to withstand extreme amounts of damage and still keep flying. But what about the general-aviation airplanes that we fly today?

In Chapter 9 I explain what happened to forty-seven door-popper, but Fig. 3-1 demonstrates much more graphically than I can possibly tell you just how much damage the airplane suffered. Controllability was certainly hairy, but it did fly, and it was controllable. It is truly amazing how much damage an airplane can sustain and still keep right on flying. Figure 3-2 shows another one. Also see Figs. 3-3 and 3-4.

**3-1** *That's the entire door stuck in the tail. Note the amount of rudder trim.*

**3-2** *The pilot hit a deer on takeoff and flew his 210 over 100 miles like this.*

**3-3** *Another view of the damaged 210. Although he felt the bump when he hit the deer, the 210 flew so well with a badly damaged horizontal stabilizer and elevator, the pilot didn't even know anything was wrong! How's that for taking abuse and punishment and still truckin' right along?*

**3-4** *A final look at the 210 after the deer strike.*

# 4

# Human factors

## I Introduction to human factors

Instructors spend countless hours teaching students to fly an airplane through each maneuver by its attitude, but not only is the attitude of the aircraft in its spatial relationship to Mother Earth important, but also of at least equal importance is the attitude of the pilot who is manipulating the aircraft through the sky. This is where the human factor enters the picture.

The vast majority of National Transportation Safety Board aircraft accident reports conclude with the statement that "pilot error" was the probable cause of the mishap. It is therefore in this area that we can effect the greatest improvement in our accident record. If we can substantially reduce the errors we make as pilots, we will make quite a dent in the number of aircraft accidents. Thus more and more attention is being devoted to human factors in aviation. It is to this area that the FAA is devoting the most attention, and rightly so.

The mental mind-set that human beings carry with them will frequently determine their reactions to whatever situation in which they find themselves. And the FAA has identified five particularly hazardous attitudes that all pilots sometimes exhibit to a greater or lesser degree. Rarely do any of these factors stand alone. Rather, they are usually found in combinations of more than a single attitude held by the same individual. For each of these five hazardous attitudes, the FAA offers us an antidote. The objective is, of course, to replace these undesirable attitudes with positive, safe mind-sets.

The five dangerous attitudes identified and defined by the FAA as hazardous are invulnerability, antiauthority, impulsivity, macho, and resignation. Let's first consider each of these very human attitudes, along with examples of each, and how they, alone or in combination, work to set the stage for disaster. Then we can consider the antidote for each and just what we as individuals can do to counteract these attitudes when we find them within ourselves. (Fig. 4-1.)

## II Invulnerability

The first step in preventing an accident is acknowledging that it can happen to me. All of us embrace, to a greater or lesser extent, the feeling that accidents happen to all those other foolish people, but never to us. After all, with our armor of knowledge and skill, we are bulletproof. Occasionally we may see, hear, or read about an accident and think to ourselves, "There but for the grace of God go I." But on the whole, deep down we don't believe it can happen to us. It is always the other pilots who skimp on the preflight and take off with contaminated fuel, or insufficient fuel, who neglect to lower their landing gear, who mismanage their fuel systems, who fall victim to "get-home-itis," or a myriad of other little things. And if each of us thinks back and does some honest soul searching, we can recall occasions when we have been guilty of these little lapses. It is not always the other people. It can be you or me! And it might very well be us if we are not careful to avoid permitting this dangerous attitude to creep in.

**4-1** *Protected by the armor of invulnerability, the pilot sets out to prove something.*

In the old flight-training manuals there was a fictional cartoon character called Mr. McFishbiscuit who was used as an example of what not to do. Here, however, we will use examples from real life to illustrate the problems that result from the hazardous attitudes we have identified. Only the names of the individuals involved have been changed. Otherwise, each example is taken from an event that actually occurred.

A pilot whom I knew quite well, we'll call him Sam, surely believed himself to be invulnerable. Sam was an absolutely outstanding airplane manipulator. He was nothing less than superb insofar as flying the airplane was concerned, but he thought he was invulnerable. He was so good that he thought nothing bad could ever happen to him. Over the more than 10 years that I knew this guy, he had several minor mishaps from which he could and should have learned better, but he steadfastly refused to learn from these experiences.

For example, in taking a shortcut to a parking spot on an airport, he taxied into a ditch, destroying a propeller and a nose strut. In that particular situation, it would have cost him less than a minute to stay on the paved taxiway. Another time he started an engine on a high-wing twin with a ladder leaning against the cowl, and pieces of aluminum ladder went flying all around the airport. On still another occasion, at night in VFR conditions, he invented a new technique for handling a gear-up landing in a twin Comanche. He simply applied power, and leaving a series of gouges in the runway, along with a belly strobe and antenna, he went around the pattern, extended the gear, and landed with two homemade Q-tip props. He then turned off the runway, taxied to his hangar, and put the airplane away, all without calling the tower or answering the controller's calls! We all know what happens to people like that. Eventually it catches up with them. Given enough exposure, they will get caught. I've seen this same guy file an IFR flight plan and charge right out into heavy weather without bothering to get a briefing, and without a chart aboard. (He was in such a hurry, he left his flight bag with all his charts on the desk.)

I need not relate his final adventure here, but I'm sure you can imagine what his final fate was. I've listed here only a few of his minor episodes. He had a great many more opportunities to learn, and if any of them had been a learning experience instead of what he considered some outside factor or bad luck, his final tragedy might well have been avoided. Sam just knew that whenever anything bad happened it was the fault of something other than poor judgment or a bad decision on his part. After all, he was invulnerable. Each of his little lapses might seem like stupidity, but let me assure you that in terms of raw intelligence, he was an extremely bright guy, a graduate engineer who had acquired several patents in the field of hydraulics. However, what he lacked was mature judgment. He believed himself to be invulnerable.

The antidote for the attitude of invulnerability lies in pilots acknowledging that it can happen to us. We must force ourselves to this realization, emotionally and intellectually. We must be careful to never put ourselves in a box with a lid on it. In other words, as the FAA puts it, "Always leave yourself an out!" Personally, I most definitely believe in this principle, and in an effort to avoid becoming complacent, every day I tell myself and force myself to really believe that indeed it can happen to me. There's nothing special about me. I'm no great shakes as a pilot, just an average guy, but I've been flying for well over 50 years and I consider myself a survivor. I'm running scared every time I get into an airplane.

Repetitive tasks can easily lead us into a false sense of security, and so it would be easy for me to become complacent. I've had more than a couple of friends of my generation who died in airplanes as a result of what I'm sure was simple complacency, so my personal first step in preventing this is forcing myself to really believe it can happen to me.

Invulnerability is closely related to complacency, and we must never allow ourselves to become complacent. No matter how familiar our equipment has become, no matter how often we have repeated the same routine tasks, we must use the checklist, really use it, and not just brush over it as many pilots do when they feel totally familiar and comfortable with the equipment. No pilot ever took off expecting to have an accident. Therefore, if we are constantly alert for the unexpected, we won't be caught by surprise, and the key to success is to make sure there are no surprises.

For three and one-half years I commuted daily between two airports via a route that took me over a densely populated area, and I had an emergency field in mind every inch of the way. These potential landing areas included a racetrack, two golf courses, and a high-school football field, among others. Even when I was IFR in IMC (in cloud without ground contact) I knew exactly where I was and where I was going if the unexpected occurred. And there were many days when I was solid on the gages all the way with approaches to minimums at both ends. Believe me, I'm just an average pilot, but I'm a cautious one. I'm running scared all the time, and I'm constantly thinking about what I'll do if such and such happens. I play the "what if" game.

If we carry with us this attitude of expecting the unexpected, there will be no surprises, and when the unexpected does happen, like the Boy Scouts, we'll be prepared. And if pilots keep their cool, there's almost nothing that can happen in a light airplane from which they cannot escape relatively uninjured, and with minimal or no damage to the machinery. I know this because in my career as a pilot, I've had three unscheduled landings in airplanes (gliders don't count), and in each case I was successful in putting the airplane down without either injury or damage. Two of the three were off-airport landings due to engine failure, and in one case I made the airport with partial power. If we just keep our cool (remember, it's panic that kills) and do as we were taught, we'll be okay. We should be confident of our ability, but we must not permit overconfidence to become complacency which leads to invulnerability.

# III Antiauthority

"Rules? We don't need no stinkin' rules!" We've all heard fellow aviators express these sentiments at one time or another, and most of us at one time or another have had cause to question one regulation or another, or have had reason to question the instructions of a controller. Why am I being vectored way over here instead of being permitted to fly a more direct route? Why am I being held way out here when there doesn't appear to be any holdup down the road? Why can't I get my release and get going? What do you mean my airplane is grounded until I comply with that new airworthiness directive? I can't see where that little discrepancy makes it unairworthy! Have a rating and get a clearance to fly in cloud? What for? I know how to keep the airplane upright and get where I'm going!

It is what we do about it rather than what we think that makes the difference. If we overtly defy the controller or deliberately violate the regulation that we feel is an unnecessary burden, we are asking for trouble, not only in the form of possible enforcement action by the friendly feds, but also in some unsafe action that could easily result in disaster.

We've all seen examples of this antiauthority attitude, or at least heard pilots verbally express their feelings against certain regulations and relate just how they got away with disregarding these regulations every chance they got. In other words, we've heard them brag about the violations they got away with. Our friend Sam, in addition to feeling invulnerable, also carried with him a substantial amount of the antiauthority attitude.

Then there was the young pilot who disregarded all the regulations. He had even been caught carrying a passenger when he had only a Student Pilot Certificate, which clearly stated "Passenger Carrying Prohibited." His excuse was that the young man he took up had helped him wash the airplane belonging to the flight school where he was training, and he owed the guy an airplane ride. This debt took precedence in his mind over the regulation forbidding him to give the ride, or any other regulation which would get in the way of what he wanted to do.

This particular pilot flaunted the rules. He refused to listen to any of the several instructors he had on his way to the private certificate. (At least three CFIs that I know of refused to attempt to train him because of his attitude.) In fact, he insisted on arguing with each of his numerous flight and ground instructors in an effort to prove them wrong on one point or another. He finally killed himself in an attempt to make an impossible takeoff and climb out. Fortunately, the passenger he had with him at the time survived relatively uninjured. This young man demonstrated at least three of our hazardous attitudes, among which antiauthority was foremost, and his final adventure was predictable, just as surely as airplanes fly. Almost everyone who knew this kid knew what his end would be. The really sad part is that nobody could tell him anything. Three separate flight schools had sent him away, refusing to work with him because of his defiant attitude toward authority.

Most of us are good, law-abiding citizens who rarely knowingly break the law, but we all harbor within ourselves a certain amount of this antiauthority attitude. And a certain amount of this is desirable, for if we blindly follow whatever we hear from those in authority and fail to question unreasonable pronouncements, we have surrendered our individuality. However, carried a bit too far, we may find ourselves embracing the antiauthority attitude which will not only get us in trouble, but also can be downright dangerous to the health of the human body.

We are all no doubt less than thrilled with some of the regulations with which we have to live, and I'm sure that each of us is sometimes unhappy with specific instructions from Air Traffic Control, but it is still a free country, and if you can't play by the rules, don't play the game! With respect to combating this antiauthority attitude, we must all realize that, mistakenly or not, the rules were made for our own good, and to break the rules is downright foolish. And nobody wants to be thought a fool!

The flight review is doubtless the most ignored regulation of all, and it is probably the only opportunity most pilots get to see just how well they are performing, to be evaluated by an impartial observer. Many pilots assume the posture that since they already demonstrated competence at the level of their certificate at the time they acquired the certificate, they shouldn't have to keep proving their ability again and again as the flight reviews come due. We're not referring here to the pilot who forgets and simply lets the review slide past, but rather to the defiant ones who refuse to submit themselves to review. Pilots who exemplify the antiauthority attitude can be heard by all who will listen as they brag about the violations they got away with. The very best thing that can happen to these people is for them to get caught and suffer a suspension or revocation (depending on the seriousness of the offense, whether it is deliberate, and whether it has been repeated). Of course, a suspension or revocation may only engender more resentment and reinforce the defiant attitude. As the FAA puts it, this guy says, "Don't tell me what to do!" And to paraphrase the FAA antidote for the antiauthority attitude, "Obey the rules. They are usually right and they are made for our own good." By the way, did you get that "usually" in there? That's the FAA talking, not me.

If a regulation is truly offensive and useless so as to serve no safety purpose, there are means by which we can get it removed, or at least changed.

# IV Impulsivity

The urge to do something immediately without deliberate consideration of all the available options is the next of the FAA identified hazardous attitudes and is called by the name of impulsivity. This is what causes pilots to make an instant response to a situation, and often it is the wrong response. Students at all levels are told, "There's nothing happening so fast that you have to panic." When the heart comes up into the throat, the adrenaline starts pumping, and the sweat starts pouring, some pilots are inclined to grab the first knob or control at hand and move it just to be doing something, anything. They experience a strong urge to do something, anything, and to do it immediately. The action they take is frequently the wrong response, and serious trouble ensues.

Here, too, we have a simple antidote, but it is likely to be difficult to apply. We must force ourselves to pause, take a deep breath, and think. And this is no easy task. The United States Navy trains its pilots, when confronted with an emergency situation, to stop whatever they are doing and wind or reset the clock on the instrument panel. This simple action requires pilots to relax. There's always time enough for this. And while thus engaged, they can review all options and make the proper selection.

The solution to overcoming impulsivity, then, is, "Not so fast. Think first." Here we are confronted with a narrow line between indecision and immediate (probably wrong) action, and of the two it is my personal belief that indecision is probably the more dangerous. We must be decisive, but only after a brief review of all the possible courses of action and the selection of an appropriate one. Remember, nothing's happening so fast that we have to panic! We must act with all deliberate speed, but the key word is "deliberate." Remember, it's panic that kills!

In order to survive an emergency, pilots must try to remain calm and apply the techniques they learned during training. There is absolutely no way pilots can be taught the correct response to every possible situation with which they may be confronted, but they can be taught how to recognize the problem, and once they've done that, they can work out a solution. The key here is to think. If pilots refuse to permit emotion to take over, they can apply reason to solve their problems and select the best available option, rather than blindly doing something, anything, in a panic. In multiengine training, applicants are taught to run the procedures with all deliberate speed when they lose an engine, and the key word here again is "deliberate." I know of at least one experienced pilot who lost an engine on climb out and was in such a hurry to do something that he feathered the wrong engine, with a disastrous result. He died in the crash. That guy was, fortunately, alone in his airplane, so at least he didn't take anybody with him. Another classic fatal in a twin-engine airplane occurred a few years ago in a training accident at my home airport. Three people lost their lives as a result of an immediate reaction to an engine cut, a wrong reaction! The lesson is clear. Don't get in a big hurry to do something.

I've never forgotten a lesson I learned from my first primary flight instructor, way back in 1942. We spent a half day in ground school and a half day on the flight line for 13 weeks, training in open cockpit airplanes with only one-way communication through a *gosport* (a speaking tube through which the instructor could yell at the student, but through which the student could not answer—not unlike a doctor's stethoscope). With the instructor in the front seat of a tandem airplane, all the student pilot could see in front was the back of the instructor's head and an occasional hand signal.

Each time my instructor would call for a maneuver, I would sit there and think through what I was going to do with my hands and feet and just how I was going to move the controls. In other words, I would do nothing. After screaming at me for a few hours, my instructor finally, in desperation, took me aside and in language not suitable for this refined publication explained that when he asked for a maneuver, he wanted me to do something. If what I did wasn't right, he could correct me and show me again just how to perform the maneuver, but if I did nothing but sit there and think about it, he would be unable to guess just what I intended to do. My problem was indecision, not failing to decide between courses of action, but not determining a course of action. From my instructor's viewpoint, this was worse than if I made a wrong decision.

The point is that in countering each of the FAA-identified hazardous attitudes, we must strike a balance. In the first instance, it's between timidity and overconfidence. Then it is between failing to question erroneous instructions or regulations and opposing authority merely for the sake of doing so. Next comes the narrow border between impulsively rushing to do something, anything, and surrendering ourselves to events, letting the airplane fly us instead of us flying the airplane. In other words, permitting the situation to take charge instead of retaining command of our own destiny. We can make a difference. And this brings us up to the next of our undesirable attitudes.

# V Resignation

The friendly feds explain resignation as the pilot thinking, "What's the use? Nothing I do will make a difference." This is the one in which the pilots surrender themselves to fate and permit whatever will happen to take place, while they merely sit there and do nothing, believing that nothing they can do will make a difference. These are people who simply give up, and rather than remaining in control of the situation, they permit the situation to control them. They have ceased to be pilots and have become passengers. They might be called fatalists, believing that forces outside themselves are in control. Their lack of reaction is, of course, exactly opposite that of the pilots who espouse impulsivity. Instead of impulsively doing the wrong thing, and doing it in a hurry, they merely sit there and do nothing at all, blithely permitting outside forces to control their destiny. As I pointed out earlier, indecision is just as deadly as wrong decision!

As an illustration of the disastrous results of the resignation attitude we have the case (a real situation) of a pilot with very little experience in the complex twin he had recently bought. He was also a relatively inexperienced instrument pilot, a dentist returning from a vacation with his entire family aboard when he flew right into the side of a mountain, killing himself and his six passengers as a result of the attitude called resignation. This pilot just let go and let fate take over. His philosophy was, "Whatever will be, will be." This is entirely different from the case of the person who freezes with terror. In that case, the emotion of fear has taken charge, and when the emotions take over, logic takes a hike.

The multiengine, instrument-rated pilot referred to here had given up much earlier in the fatal flight. When a problem first manifested itself, he had turned complete control of the flight over to air traffic control. Then, when the controllers lost him on the scope, he just gave up completely and let fate take over while he blithely flew into the mountainside.

A somewhat different but related phenomena is a case I had when attempting to train a man of the cloth. He had been a fair student until we started serious work on landings in the traffic pattern. Every time he got the airplane somewhere near the runway, he would just let go and I would have to take over and land the airplane. When I asked him just what he had in mind, he informed me that he'd gotten us this far, now God could take over and land the airplane! (Fig. 4-2.)

I tried to explain that while the good Lord might be right there with us, it was up to the pilot to manipulate the airplane, and that no

**4-2** *I've done all I can; now it's up to the Lord.*

doubt it was God's intention for the pilot to do the flying. Although this happened many years ago, to this day I'm not completely convinced that I was successful in getting that idea across. While not a real example of the resignation attitude at work, this case exemplifies the same attitude that leads pilots to say to themselves, "I can't do it, so I'll let some other, greater power take over."

Pilots who resign have surrendered any control they may have had over the situation. To counter this proclivity to give up and resign oneself to the fates, it is only necessary for pilots to put their minds in gear, think through their options, and take appropriate action. The only thing required is for pilots to realize that they're not helpless, and as the FAA puts it, the pilots can make a difference.

## VI Macho man (or woman)

Everyone knows more than one Mr. (or Ms.) Macho, the tough people who are not only confident but overconfident. Given a choice, they will take the risky over the sure, safe course. These people are sure they can handle any situation in which they have put themselves, and they know no fear. Each successful adventure reinforces their self-image as superhuman until one day they attempt the really impossible and disaster befalls them. They're "kick-the-tires-and-start-the-fires" aviators who can't be bothered with a thorough preflight inspection of the equipment, and who don't require a chart because they know where they're going and how to get there. They don't need a weather briefing either. After all, they can look out the window and see what the weather is doing. Who needs a briefer to tell them what they can see with their own eyes? And besides, they're sure that they know more than the briefer, who probably isn't even a pilot anyway. They think checklists are for wimps. These macho people know all the different airplanes they fly and how to work their systems and special equipment. In fact, they know more about aviation than anyone around them, and if you don't believe it, just ask them! What they fail to realize is that it is much better to learn from the mistakes of others than from their own. Our friend Sam, whom we met back when we were discussing invulnerability, is a prime example of Mr. Macho Man. (Fig. 4-3.)

We all know several examples of people with the macho attitude. They're the people who set out to prove that they can do the impossible, and when they discover that it really is impossible, it's too late! All pilots must be confident of their ability, but they must not be overconfident. We should also remain aware of the fact that a lack of confidence can be equally bad.

**4-3** *The macho pilot can even get away with flying under the wires.*

Here again, when pilots get away with foolish, dangerous acts, their macho attitudes are reinforced until the time comes when they either scare themselves out of it or encounter a situation that is quite beyond their ability. I recall a pilot who was an habitual scud-runner, constantly bragging about how he could always sneak in under a low ceiling. One day he was scud running home and when he finally spotted his home field, he had to duck under the wires that cross the road on the final approach to the runway in order to get in and land. That episode shook him up to the extent that he never again attempted to fly in marginal weather. He finally acquired an instrument rating, an effort inspired no doubt at least in part by his scary experience with the power lines. Fortunately, that pilot learned from his experience, but it could just as easily have turned into the material of which newspaper headlines are made. Of course, had his macho attitude been strong enough, he wouldn't have learned from the experience and would have kept going as he was until he really got bitten.

The FAA says the antidote for this one is to realize and tell yourself, "Taking chances is foolish." If we make ourselves face this fact and substitute this thought for the "I can do it" attitude, we will have counteracted the macho attitude, and one more of these dangerous mind-sets will have been removed. I knew a student pilot several years ago who read a book on aerobatics, took out a rental airplane without the authorization of his instructor (he was only endorsed for solo in the pattern), and attempted several aerobatic maneuvers. He was lucky. He severely overstressed the poor little Cessna 150. He brought it back with stretched control cables, and bent ailerons, wings, and empennage, but he did bring it back and effect a safe landing. However, in this case the experience only served to reinforce the guy's sense of machismo and invulnerability.

Another student pilot wasn't so lucky. This one, on his first solo flight, disregarded his CFI's instructions to make three landings, remaining in the pattern, and return to the ramp. He made two landings and then departed the pattern, flew over to his home and buzzed his house at treetop level so his wife could photograph the event. He then flew over to his daughter's home, buzzed that one as well, and on the pull-up from a low pass, he stalled and crashed. This man's macho attitude cost him his life! And his estate had the effrontery to sue the flight school where he had been training. They brought a suit for wrongful death on the theory that he had been getting inadequate training.

This one is an interesting case because it includes all our hazardous attitudes with the sole exception of resignation. The guy who did all this was a very prominent, early-fifties businessman who lived in a rural community in the Midwest. He owned a pressurized twin which was flown for him by a professional pilot. The owner of this airplane was the leading citizen in his hometown, which is the county seat. Located there is a small county-owned airport with one paved runway. The individual being discussed here owned a manufacturing plant which was the town's and the county's largest employer. He was not too well liked in the community since he had a reputation for being overbearing and throwing his weight around. All in all, he presented a very intimidating persona.

This high-powered guy had a second home in Florida to which he regularly flew in his professionally operated pressurized twin. However, he had started taking flight instruction in a two-place trainer at the county airport near his home in the Midwest. On occasion, when in Florida, he and his employee pilot (a certificated flight instructor) would go to a local FBO and rent a four-place airplane in which the instructor would add to the businessman student's training, and when they traveled in the twin, the professional pilot would sometimes allow his boss to twist the yoke and push the pedals. Altogether, with his training at home up north and in Florida, the high-powered businessman pilot had about 30 hours of instruction but had not yet soloed when his career as a pilot ended. His training had been spread out over more than a year's time.

One fine late-spring day, he showed up at the county airport where he had been training, announcing that here he was, ready to go, and demanding an instructor to serve him. This was his habit. No appointment, just show up and demand service. He was used to getting his way in all his relationships, so why not with respect to his flight training? The FBO had one full-time instructor on staff, and as it happened the instructor was available. The instructor was fairly new, both in terms of employment at that FBO and as a certificated CFI, and he had not flown with this student previously.

The student, a very intimidating guy to say the least, informed the instructor that they would go out and do landings, implying that he was about ready to solo and was anxious to get on with it. The instructor was obviously somewhat overwhelmed by this heavyweight with the big reputation as a highly successful businessman, with a not inconsiderable amount of influence in the community. Even so, he reviewed the student's logbook and noted that all the required presolo maneuvers had been signed off by either the previous instructor at the FBO or by the guy's professional employee twin-engine pilot. This being the case, the instructor agreed to take the guy out and work landing practice.

They went out and flew the pattern, doing takeoffs and landings for about an hour and a half when the instructor, under pressure from the student, offered to solo the guy.

"Let me have your logbook and your student certificate so I can sign you off. Then go out and make three landings for me," he said.

The student informed the instructor that he didn't have his student pilot certificate with him. He had left it in Florida, "But I'll get it," he said.

"In that case, come back when you've got it," was the reply.

"Okay," said the student after some argument, and he left.

A few days later, again without an appointment, he showed up with his student pilot certificate, which had been sent to him by his Florida housekeeper. This time, however, the instructor was working with another student, planning a cross-country training flight. The big-shot student demanded immediate attention as was his habit, even though he could see that the young instructor with limited experience was busy with another student at the time.

Conditions were absolutely perfect, so the instructor excused himself briefly from the cross-country planning session to once again check that the domineering student had met all the presolo requirements (stalls, emergencies, etc.). He then signed him off and dispatched him with firm instructions to make two touch-and-go landings and a full-stop landing and return to the airport office, anticipating that by the time this was accomplished he would be free to attend to him. This all took place in front of three witnesses.

Now there's nothing in the regulations prohibiting this procedure of dispatching a student on an initial solo without flying with him on that same day, but it is certainly unusual. The normal procedure is for the instructor and student to be doing landings together, and after several unassisted good landings, the instructor will let the student go for three by himself. Even so, under the existing circumstances it is quite understandable that the young instructor, who admits to having been intimidated by the student pilot, would send him out by himself without having flown with him that day.

I've already related what happened next. The big shot's fatal buzz job not only resulted in his demise, but in the total destruction of a perfectly good airplane as well. The guy had even told his wife before he left for the airport that he expected to solo that day, and she should go out onto the upstairs porch with a camera so she could photograph the event. Indeed, she did get some pictures, but they weren't what he had expected.

The individual in this case displayed extreme examples, a truly classic case, of the attitudes of invulnerability: he just knows it can't happen to him (antiauthority); he consistently defies the rules (impulsivity), in a hurry to do it right now; and macho, all combined in one individual. It is often so. Put this deadly combination together, and the result is bound to be disaster. The same person will carry with him or her more than one of our hazardous attitudes. This one is another example of a fatal crash (we can't very well call it an accident) that was predictable with absolute certainty, given the attitude of the student. You can see these coming almost from the time you first encounter the individual. Given enough exposure, any so-called pilots with that kind of attitude will get bitten. If they're lucky, they'll scare themselves so badly that they'll change their entire attitude toward flying, but that's really the only way such people can be made to change. If they have one really close call, perhaps they will then realize that indeed it can happen to them.

There's another lesson in this story as well. It matters not how much older, how much better educated, how much more money the student makes; in the student-instructor relationship, the instructor is the expert. And this fact must be impressed on every single student with whom an instructor works. Instructors simply cannot permit themselves to be dominated by students and permit students to call the shots with respect to their training. The instructor is the expert in the field.

## VII  Hazardous attitude summary

There are times when all pilots make a choice based on taking a calculated risk. We take this action as a result of having weighed a series of options and selecting one which perhaps involves some small amount of risk. We minimize the risk factor and we call this "taking a calculated risk." This is a completely different thing from taking chances. That involves the habitual taking of unnecessary risks. Macho pilots are overconfident. They just know they're good, so the normal safety precautions don't apply to them. They persistently take chances.

We all need the confidence to fly to the limit of our capabilities, but not beyond the limits. And there's a thin, hazy line between confidence, which is desirable, and overconfidence, which is definitely not. As pilots we must know not only our capabilities, but also our limits. And this applies not only for us as pilots, but also for the machinery which we operate. The very worst thing that can happen to pilots early on in their careers is for them to get away with a series of little careless actions, for this will lead them to the belief that they can go on doing so until they, too, experience the hard lessons, or worse, until they encounter the ultimate, absolute disaster.

Personally, I like to learn the easy way. It not only takes a lot less effort on the part of a lazy guy like me, but it's also less expensive. We simply must respect the medium in which we operate, the machinery with which we do it, and our own limitations. This is not to say that we must be timid, but rather we must temper confidence with caution because being timid is as bad as being overconfident.

The exact opposite of impulsivity is the attitude identified as resignation. Doing nothing is equally as dangerous as rushing out and doing the wrong thing just to be doing something. Indecision is at least as bad as wrong decision. Pilots who find themselves in a tight spot, simply resign themselves to the situation, and let what will happen take place have committed serious mistakes. This is surrendering, giving up. We're not referring here to people who freeze with panic and are unable to act, but rather the people who are capable of action but who consciously refuse to act. On the other hand, people who freeze on the controls have permitted fear to rule their actions (or inaction).

The desired course is a balance somewhere between hasty, ill-considered action, in such a hurry to do something that we do the wrong thing, and deciding to not do anything because of a feeling that no matter what we do it will make no difference. When the situation calls for decisive action, we must weigh our options, choose a course of action, and then act decisively.

The point is that in countering each of the FAA-identified hazardous attitudes, we must strike a balance. In the first instance it is between being timid and having too much confidence. Then it is between failing to question erroneous instructions or regulations and opposing authority merely for the sake of doing so. Next comes the narrow border between impulsivity, rushing to do something, anything, and surrendering ourselves to events, letting the airplane fly us instead of the other way around. In other words, permitting the situation to take charge instead of retaining command of our own destiny. We can make a difference.

Finally, we must avoid the childish desire to show off just how good we really are. This goes back to the matter of confidence. It is almost as if the macho people are challenging themselves to exceed their limitations. The files of the National Transportation Board are replete with case history, accident report after accident report, directly attributable to the fact that the pilot had one or more of these hazardous attitudes. And it is frequently more than one. Cocky individuals who are overconfident of their abilities are also likely to hold antiauthority feelings as well, not to mention a sense of invulnerability, and they may very well also be impulsive. Put all this together, and disaster is sure to follow. It is almost never just a single item listed in the NTSB probable-cause column.

Most, if not all, of us carry within ourselves, to a greater or lesser extent, all of these attitudes. And we must all guard against letting one or more of them take charge in an emergency. When we look at the accident reports, we see time and time again examples of pilots who exhibited not one, but two, three, or more of these hazardous attitudes, which combined to lead to disaster. Let's each strive to apply the antidotes and avoid permitting any of these dangerous mind-sets to lead us into serious trouble.

# 5

# Thinking flying

## I It can happen to me

The FAA first experimented with the Accident Prevention Program by assigning an aviation safety inspector (general aviation operations type) in each of two General Aviation District Offices to the duty of establishing a program designed to prevent accidents. After the trial period, the agency firmed up the concept that the first step in preventing an accident is to acknowledge that it can happen to me. The FAA also mandated the official national Accident Prevention Program in every Flight Standards District Office by designating an inspector in each office as Accident Prevention Specialist (or APS as they were then called). However, as the accident prevention program has grown in importance within the Flight Standards division of the FAA, a movement away from the emphasis on this all-important first step has occurred, a situation to be regretted.

Safety is indeed no accident. This simply means that safety must be intentionally pursued. And in this pursuit we must first acknowledge that it is not always the other pilot who is involved in an accident; it can happen to me! This translates into action in the form of requiring pilots to be constantly alert and to expect the unexpected. There should be no surprises. If pilots are well trained and are not caught by surprise, the probability is that they will be able to cope with almost anything that comes up. But first they must face up to the fact that an accident can happen to them.

The intentional pursuit of safety requires that pilots put their brains in gear and think at each step of the way on each and every flight. If, on every single takeoff, as they apply power and start down the runway, the pilots are expecting to lose their engines, they will be prepared and in a position to take appropriate action.

They will be in control rather than permitting the airplane to control them. By being prepared and anticipating a problem, the pilots can have a course of action in mind and prevent a potential disaster.

There is almost no emergency situation in which pilots of average general-aviation airplanes cannot cope if they are properly trained and if they keep their cool. It is panic which is the killer. If pilots remain calm and do what they have been taught, they will most likely extricate themselves from the situation and avoid disaster. This, of course, is not easy. They must force themselves to keep calm, but it can be done.

Safety in general is a very difficult subject in which to get people interested. This is because accidents always happen to those other people, but if we realize emotionally and intellectually that it can happen to me, we will have taken that all important first step in preventing an accident. It is easy enough to know something like this intellectually, but it is frequently extremely difficult to realize it emotionally as well. Even so, this is a step which all pilots must take if they are to operate safely in the airspace. Our training prepares us to deal with most emergencies, but when the emotions, especially fear and panic take over, logic and reason go out the window. People have been known to become paralyzed with fear to the extent that they are absolutely helpless and unable to cope.

We are now starting to train pilots to recognize and counteract the more common mind-sets which create dangerous situations. If we can recognize and acknowledge within ourselves the underlying predisposition for a dangerous attitude, and we know and apply the antidote, we will have gone a long way toward preventing a potential accident.

There are certain conditions and situations which no amount of forethought or expectation can prevent, such as mechanical problems of one sort or another; unexpected, unforecast weather; etc. However, there are other factors which may lead to potential disaster about which we can do something. Among these are the hazardous attitudes within ourselves identified, along with their antidotes, and spelled out in Chapter 4.

## II  The system

Human beings are not designed for flight. Only most birds, insects, and a few furry mammals are designed by nature for flight. For them, flight is a natural activity, but for us it is unnatural. When we operate in airspace, we are in a foreign medium. Basically, human beings are land animals. Although we can swim in the water under our own power, in order to fly we require a machine that was invented by human intelligence. Since this is the case, we must apply intelligence to the activity of flying. It simply does not come naturally.

Because flying is not a natural human activity, everything about flying must be learned. The physical acts involved in manipulating an airplane through the sky consist of a series of conditioned responses. Thus, to be successful, flying requires repetitive training and consistent practice. In order to build a conditioned response, an activity must be repeated over and over again until it becomes "second nature," the natural thing to do. This is why flight training requires that each maneuver be repeated until the students have it down perfectly—repetition and drill, repetition and drill. And then when they have got it down, they have to keep doing it fairly frequently or they'll lose it. The old "use-it-or-lose-it" syndrome comes into play. This factor is recognized by the FAA in the requirement for recurrent training for those who carry passengers and property for hire, and the flight review for those who don't.

Since time immemorial humans have envied the birds as they soar through the air. However, nothing about flying is natural for human beings; it all must be learned. Through constant practice and repetition we build a set of conditioned responses which enable us to operate in the unnatural environment in which aviation takes place. The correct responses come only through repetition and drill until they become "second nature." And then they must be reinforced with frequent practice. We call this "staying current."

And this business of staying current is one area that is frequently ignored by a great many pilots. The people who earn their living by flying commercially are required by law to stay current and prove it with frequent flight checks, but the average people who use aircraft for personal business and pleasure need only perform the minimum of three takeoffs and landings every 90 days if they want to carry passengers, and they must demonstrate currency in knowledge and skill annually or biannually to an instructor.

Dr. Gustavo E. Cosenza, who is a Director of the Board of the Aeroclub of Guatemala and Coordinator of Education and Safety Commission, in his presentation to the winter meeting of the Lawyer

Pilots Bar Association in 1993 on the subject of Neurology of Flight, not surprisingly likened the human brain to a computer with a central processing unit. He pointed out that pilots are gadget lovers, what with GPSs, Loran Cs, CRT displays, Moving Maps, Flight Directors, HUDs, etc. Still, the most important computer on the airplane is the one between the pilot's own ears. Statistics indicate that some 85 percent of aviation accidents result from what is called pilot error, the failure of the CNS (central nervous system) to respond in a proper and timely fashion to the complex airplane-weather-pilot-ATC system interaction. It is in this area that the most drastic improvement can be made. And if we are to do better, it is first necessary that we have an understanding of how that system works.

Dr. Cosenza is a neurologist; Fellow of the American Academy of Neurology; designated a Diplomate in Neurology by the American Board of Neurology and Psychiatry; a certified flight, ground, and instrument instructor; and holds an aircraft mechanic certificate for airframes and power plants. He defines the human nervous system as a computer having hardware (input-CPU-output-feedback systems) and software which include what the pilot is, knows, and can do. The input is given by the visual system, the vestibular system, and the somatosensory system (seat of the pants). The CPU (central processing unit) includes the processor capacity (IQ), RAM and ROM, logic, math, visuomotor, and spatial orientation units. The output is supplied by the neuromuscular system, and feedback is given by the extremely fast interaction of these systems.

The software implanted in the brains of pilots includes their training, which consists of an abundant supply of subliminal inputs from the training environment (instructor and peer attitudes and ground and flight instruction and practice); their proficiency, including regular recurrent training, both formal and informal, which is equally as important as initial training (the "use-it-or-lose-it" concept again); their experience, for which there is no substitute and which must come gradually (one cannot force it by attempting to "cram" it in. It includes the influence of the flight environment and role models and post-flight self-debriefing.); pilot judgment, which is the result of all these components and which is both subtle and hard to define (this is the mark of the good, safe pilot.); and psychological makeup, which may determine the pilots' responses in stressful situations. When the emotions take over, logic and reason go out the window, and fear is an extremely powerful emotion. As Dr. Cosenza puts it, "Fear is the mind killer."

Dr. Cosenza then discusses the brain as a biological system, again comparing it to a computer. He explains cell organization and function, the neurons, and how they function and communicate. His slide presentation shows how the electrochemical impulses charge through the synapses from one brain cell to another with absolutely incredible speed. He explains the needs of the neurons (oxygen and energy), and how they network. He also points out that exercise improves both their function and structure.

Dr. Cosenza describes the functional organization of the central nervous system and operation of its subsystems as they relate to piloting. These subsystems include vision (possibly most important), vestibular, somesthetic, the cerebellum, and the cerebral cortex, all of which combine to enable us to fly.

# III How to cope

Since the vast majority of aviation accidents result from what the FAA and the NTSB are pleased to call "pilot error," it would certainly seem that our efforts to reduce the accident rate would be better directed here than elsewhere. Of course, we want the machinery to be as safe as it possibly can be, but we already have perfectly safe machinery and we still have accidents. The machinery simply can't be made foolproof. Just as soon as we think we've made something foolproof, something that can't possibly be screwed up, some idiot comes along and proves us wrong. Consequently, it behooves us to turn inward in our search to improve the safety record.

There are numerous things pilots can do to ensure that their CNS (central nervous systems) remain in an airworthy condition. Assuming the pilots have, under FAR 67.13 d, met the mental and neurological requirements of certification, and have complied with Chapter 8 Section 1 of the AIM, Dr. Cosenza points out that pilots can help to keep themselves mentally fit by applying Grandma's recipes of nutrition, rest, exercise, sleep, and above all, maintaining a good attitude.

Of course, pilots must work at keeping toxins out of the cockpit, including alcohol, smoking, medication, carbon dioxide, etc., and pilots must use oxygen as appropriate. They must also carefully manage their emotional status. This includes the avoidance of stressful situations (a good idea at any time, but especially important when flying), depression, and they must not permit distractions to take their attention from the job at hand—flying the airplane. And this is not always easy, especially on a long trip when everything seems to be working okay and the airplane is just droning along. It becomes easy to be overcome with a sense of euphoria, particularly if one is at or near an altitude where oxygen might be appropriate.

Pilots must remain active and current with their skills, keeping themselves informed. In other words, they must stay well ahead of the airplane and the flight situation at all times. It is important for pilots to have confidence in their skills and training, but they must avoid becoming overconfident. You know, the "it-can't-happen-to-me" attitude. Pilots must be aware of the fatigue factor and manage it, seeing to it that they have an adequate and comfortable flight environment. They must be able to recognize and counter the effects of vertigo and the numerous illusions of flight. Pilots must take care of themselves and their equipment. They must make themselves believe that "it can happen to me."

Altogether, pilots would do well to preflight themselves and the aircraft by going through the "IMSAFE" pilot's checklist:

- Illness
- Medication
- Stress
- Fatigue
- Alcohol
- Emotion

Every pilot is familiar with this personal checklist, but what does it really mean? Let's now examine each of the items on this list.

Illness refers to virtually any condition that is less than perfect health. Even the common cold is a particularly undesirable condition in which many pilots attempt to operate. A stuffed-up head can disturb the vestibular apparatus of the inner ear, resulting in vertigo and loss of equilibrium, causing loss of control and damage to the machinery and unfortunate effects to the human body. Pilots should be aware that substantially everything other than perfect health entails a negative consequence on their ability to successfully manipulate an airplane.

I recently heard an interesting theory advanced by a knowledgeable pilot, who shall remain anonymous. He poses a real medical quandary. Our system is such that pilots seeking treatment or advice about their health are legally required to report this fact to the FAA. Therefore, according to my source, there is incentive for pilots to avoid doctors because anything that can come to the attention of the FAA can do pilots no good whatever. They might be grounded! Pilots therefore are less healthy than the population in general. My informant says this system definitely needs fixing.

Medication is another subject which needs to be addressed. The surgeon general tells us that practically the only acceptable medicine for pilots to use when they are expecting to fly within the next several hours is aspirin, and he is not even sure about that. Antihistamines are especially dangerous because they are so easily available over the counter, and many of them cause drowsiness and a general slowing down of thought processes and reaction time.

Stress causes the activation of the sympathetic nervous system, the so-called "fight-or-flight" response in the human being. This may result in the inability of pilots to weigh their options with consideration. In other words, it can have a negative effect on the decision-making process. However, like many other factors, there are both good and bad elements involved in stress. A small or moderate amount of stress tends to heighten our awareness, and thus our ability to solve problems and cope with unexpected situations. It helps us rise up to meet a challenge, just as a student gets "up" for a test.

The very nature of aviation induces a higher level of stress on a person than most other human activities. Pilots who cope with the stresses of both normal life and those which go along with the operation of an aircraft are certain to be mentally challenged even under normal circumstances This may very well be one reason that aviation has always been viewed by outsiders as a rather high-risk activity. In order to successfully cope with stress, we must recognize stress and understand stress management and the relationship between stress and performance. Unlike many other activities in which human beings engage, aviation demands a high level of performance, and consequently a certain fitness for the operation of an aircraft. Under many, or even most, conditions of flight, the required performance is easily maintained, particularly when the pilot is confident, knowledgeable, and in good physical, mental, and emotional condition.

Reducing the pilot's performance level through illness, stress, medication, or fatigue reduces the margin of safety. While most of us easily recognize and acknowledge the risks of flying while impaired by alcohol, illness, or medication, it is much more difficult to assess one's level of impairment as a result of stress, which affects performance in a much more insidious manner.

Stress, while normally viewed as a negative, is absolutely necessary in the performance of crucial tasks. A pilot's performance of crucial tasks actually improves under a moderate level of stress. One becomes psyched up for the challenge. At the low end, a pilot who has absolutely no stress, is under no pressure in a crucial situation could easily act complacently or indecisively. At the opposite extreme, a pilot who is overloaded with stress and under extreme pressure may act impulsively or break down and be unable to perform at all. This is what I mean when I say you must keep your cool and remain in total control of the situation. See Fig. 5.1, which depicts the hypothetical relationship between stress and performance. A crucial level of stress (arousal) proves optimum for performing crucial tasks, but either too little or too much stress results in inadequate performance. As Dirty Harry Callahan so emphatically puts it, "A man's got to know his limitations!"

Knowing one's own acceptable level of stress for flying is very difficult since no two pilots are alike, and there are no quantitative measures which can be used. Even so, there are a few reliable methods for determining our own unacceptable levels. First, if it just doesn't

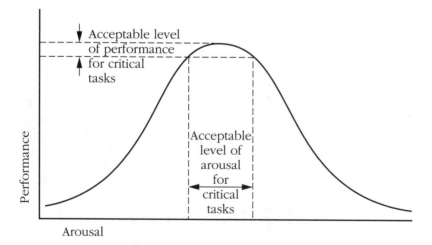

**5-1** *A hypothetical relationship between arousal and performance (after Yerkes et al., 1908).*

feel right to go flying because of the pressure under which we exist at the moment, that instinct should definitely be followed. It is usually right. Perhaps somebody is trying to tell us something. Another indicator is any major life event or series of events in short succession which tend to elevate our normal stress level. Often the elevated stress level will manifest itself in uncharacteristic behavior patterns which others notice and call to our attention. These should be factors in any decision to fly.

Corporate America recognizes that stress can cause people to be induced to react in an inappropriate manner. For example, many large corporations refuse to consider for employment an individual who has undergone a divorce within the previous two years. More than one attempt has been made to devise a scale assigning point values to such factors as marriage, marital problems, financial difficulties, etc., and the individual who rings up a high point score is classified as a greater-than-normal accident risk. Of course, it is virtually impossible to quantify this sort of thing because people are individuals, and each of us has his or her own threshold of stress.

The margin of safety for flight exists somewhere between the pilot's performance level and the workload being imposed on the pilot. When the pilot is at peak performance and the workload is low, a greater margin of safety exists for a given operation. However, under a more difficult workload, or when a pilot is impaired, the margin of safety may be reduced to the point that a very great accident risk exists. See Fig. 5-2, which depicts the hypothetical relationship between performance and workload, demonstrating just how performance can be degraded by such things as illness, medication, stress, or fatigue. The pilot's workload also increases when he or she is faced with bad weather, system malfunctions, or complex situations. Additionally, the workload is greater for landing than for takeoff, and the safety margin may be reduced to an unacceptable level.

The United States government is very big on the use of acronyms, and the military is no exception. To help get rid of self-imposed stresses, military pilots are taught the acronym DEATH, which stands for drugs, exhaustion, alcohol, tobacco, and hypoglycemia. Each of these has a direct relation to how stress affects aviators.

Alcohol is the only one of the items on the IMSAFE list which is addressed by the Federal Aviation Regulations specifically in terms of

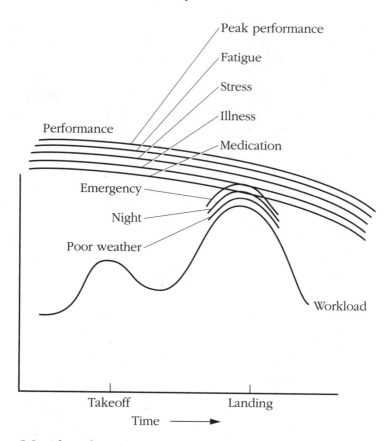

**5-2** *A hypothetical relationship between performance and workload.*

what a pilot is permitted to do. In most cases the "eight-hours-from-bottle-to-throttle" rule is woefully inadequate. Depending on body weight, general health, and stomach content it may be enough, but for the individual of moderate body weight whose stomach is empty when imbibing, it is certainly not enough. Pilots who drink even a moderate amount without a substantial time lapse prior to flying are asking for trouble.

Fatigue causes a diminishing of our cognitive ability, of our ability to coordinate, and in general leads us to become confused. In extreme cases fatigue can cause hallucination. It adversely affects our judgment, impairing our decision-making ability, which leads to wrong decisions, or equally bad, no decision at all, or indecision. Our reactions tend to become unpredictable. (See the section entitled "Confession time" in this chapter for a graphic example of this.)

Emotion, the last item on our personal checklist, may very well be the most important of all. When emotion takes over, judgment goes out the window. When things go wrong, a person is likely to become frustrated, at which point the frustration-aggression complex cuts in, again resulting in inappropriate decisions. Fear is one of the most powerful emotions of all and often results in an undesirable conclusion.

If pilots conscientiously follows this personal checklist before every flight, they will have gone a long way toward eliminating many of the factors that can lead to disaster.

## IV Confession time

I had a personal experience with the deleterious effects of the fatigue factor not too long ago. Until that night in mid-December, I had always for the most part ignored everything I had heard about or read concerning the effects of the fatigue factor on a pilot's performance, at least on my own ability to fly an airplane when I was tired almost to the point of exhaustion. I remember a day some 30 years ago when I put in over 20 hours, 11 or 12 in the air, and I seemed to handle it okay (perhaps my judgment was somewhat clouded). In this regard, and in this regard only, I have been a self-imposed victim of the hazardous attitude called "invulnerability." Intellectually, I always knew it could happen to me. But emotionally I didn't believe it could happen to me. I believed I was invulnerable to the effects of fatigue. I was in for a rude awakening!

I have been lecturing to aviation groups on the subject of the hazardous attitudes that get pilots into deep trouble for several years, and I certainly should have known better than to allow myself to get into the position in which I found myself on that mid-December night, but I simply ignored what I knew and this, coupled with that old bugaboo, "get-home-itis" almost got both me and a close friend seriously killed.

After only a couple of hours of sleep on Wednesday night, Thursday morning at 8:30 I flew a friend in a Seneca II on a 185-mile trip so that he could conduct business at one of his branch offices. It was an easy IFR trip in the morning, and when we returned that night at about 10:00 o'clock the trip itself was again easy IFR, smooth air and between layers most of the way, but the approach back home was to ILS minimums with an indefinite ceiling 200 and RVR of less than

5,000. By the time we started the approach at about 11:15, the weather had worsened slightly. I was totally exhausted. Before we had departed, I had suggested that we check into a hotel and return in the morning, but Milt had an early meeting back home the next morning and wanted to get back that night.

My friend, Milt, is a qualified multiengine, instrument-rated pilot who owns and regularly flies a Cessna 310, and although he had a moderate amount of experience in Senecas, he didn't feel comfortable flying the Seneca in weather. In fact, that's why I was accompanying him. Even so, had he not been with me, I doubt if I would be alive today. When center handed us off to approach control, the new controller informed us that in the previous hour several had made it in, but a few had missed the approach and gone elsewhere.

I listened as the second airplane ahead of us missed the approach. He broke out at 200 feet, but slightly off to the side of the runway. I continued to listen as the next one made it. Then the guy who had missed got vectored around for another attempt. He made it in on the second approach, and then it was my turn.

The problem started when I got vectored onto the localizer right at the outer marker, 300 feet higher than the published intercept altitude. The approach controller, who had his hands full working two busy airports that night, had held me high, neglecting to let me down outside the marker.

I had already retarded the power to the appropriate approach setting to slow the airplane down, and as I crossed the marker and joined the localizer, I popped the gear, applied approach flaps, and re-trimmed the airplane just as I teach my students to do. I then plunged down to intercept the glideslope, arresting the descent when I was one dot high on the glideslope indicator. I held the needle in that position, thinking that being a single dot high would be okay, just so long as I wasn't low. I concentrated on flying the gages, waiting for decision height before looking out. Milt called out that he thought he caught a glimpse of the approach lights through the thinning undercast as we passed over them, but when we finally broke out, we were over the centerline of the runway, but two-thirds of the way down the 6,000-foot runway.

There was just no way I could get rid of 200 feet, land, and get the airplane stopped and still be on the pavement, so I called the miss, cleaned up the airplane, and accepted vectors around for another try. I was again turned on to the localizer right at the marker, but this time at the proper altitude. I pulled the power back but neglected to lower the gear, extend approach flaps, and retrim. This lapse on my part was no doubt due to my being so tired that my mental functions weren't working normally.

After Milt reminded me to extend the landing gear and apply approach flaps, I held the localizer and glideslope needles both in the doughnut until just prior to reaching decision height, at which time I permitted the localizer needle to move a couple of dots to the left. On this approach we were parallel to the running rabbit, but too far off to the side to maneuver over the runway for a safe landing, so I again called the miss and applied climb power. But this time I failed to clean up the airplane. I simply neglected to raise the gear and flaps. The airplane was trimmed for landing, so the application of power caused it to pitch up sharply, and it started to climb rather steeply. In fact, it stood on its tail and headed straight up! This put us right back in the murk. I shoved down on the yoke in an attempt to reduce the steep pitch up attitude, while Milt reminded me to raise the gear and manual flaps. While occupied with this task, leaning over to lower the flaps, for the first time in my 50-odd-year career as a pilot, I completely lost control of the airplane, and when I glanced at the attitude gyro, it showed us in a 60-degree bank and diving toward the ground with climb power! Milt said he caught a glimpse of lights on the ground, which was only 200-feet below the cloud deck in which we were flying. I pulled back hard on the yoke and the airplane again stood on its tail and started up, still in the steep bank. The stall-warning horn started to sound as I leveled off and got the wings level. We were now on a heading of about 30 degrees and between 2,500 and 3,000 feet, although the missed approach procedure for runway nine calls for a climb to 3,000 straight ahead on the runway heading of 92 degrees.

Now that the airplane was back under control, I contacted approach control and advised the controller that I would try one more time, and if I didn't make it in, I would go elsewhere. (There was an airport with an ILS approach about 30 miles away that was reporting a 500-foot ceiling and 1-mile visibility.) So we were vectored around again for one last attempt.

On this third try we intercepted the localizer a couple of miles outside the marker at the proper intercept altitude, with the airplane slowed to approach speed. At the marker I popped the gear, extended approach flaps, and trimmed for the approach. I again concentrated on flying the ILS to minimums, and I held both needles in the doughnut while Milt looked for the approach lights. This time we broke out over the running rabbit with the runway right in front of us and runway lights on both sides of us. I then extended full flaps and proceeded to make a normal landing, although I can tell you my heart was pounding and the adrenaline and sweat were flowing.

What lesson can be derived from this flight, which almost ended in disaster? First and foremost, when I run my personal IMSAFE preflight checklist, I guarantee I'll never again ignore the "F" for fatigue. Of course, there was an element of that old bugaboo, "get-home-itis" at work on Milt's part, a factor which I had never previously permitted to interfere with my usually good judgment. I also promise I will never again let this happen.

I am confident enough of my skill to frequently accept tight turn-ons to final on an instrument approach. In fact, to save time I will often request a tight turn-on, but usually the weather is enough above minimums to allow for adjustment after you break out. From now on, however, when the weather is at or near minimums, I intend to insist on a reasonably far-out interception of the final approach course, and at the proper altitude. The first step in preventing an accident is to know and acknowledge that it can happen to me! Well, believe me, I do know and am abundantly willing to acknowledge that fact! Besides, I would rather be lucky than skillful any day.

# V Cockpit resource management

Recently the FAA has been emphasizing the need for CRM training for all pilots. And when the friendly feds started requiring private applicants to learn a bit about cockpit resource management, and be prepared to discuss this knowledge on the practical test, I heard a bunch of comments from flight instructors to the effect, "Yeah, I'll teach all my primary students to carefully brief their copilots in that CE 152 heavy they're flying!" Sounds pretty silly, doesn't it? But let's stop and consider the implications of this for a moment.

Any time there are two pilots in an airplane, any two pilots, it is absolutely essential that there be a crystal-clear understanding of just who will be doing what on the flight. Thus we can avoid the, "I thought you had the airplane." "No, no, you were supposed to be flying. I'm just a passenger."

Although CRM is commonly thought of as coordination between or among multipilot crews, in its larger sense it is much more than that. At the next level it encompasses the interpersonal relationship between pilots and passenger(s) and between pilots and the Air Traffic Control personnel with whom pilots deal. And finally, it can be expanded to include the relationship between pilots and the equipment they manage. In this regard you should know there's absolutely nothing wrong with talking to yourself as you go about performing your cockpit chores. You'll probably get intelligent conversation, and maybe even correct answers to the questions you ask yourself.

If the air-carrier cabin attendants can conduct a thorough passenger briefing on every flight, so can the general-aviation pilot who has passengers aboard, whether or not it is a commercial operation such as a Part 135 passenger charter. By the way, I wonder just why cabin attendants always include water survival information in their briefing for flights that are entirely over land?

Solo pilots who check and recheck the frequency and identification of the navaids they're using are engaging in cockpit resource management, as are the solo pilots who remind themselves to switch fuel tanks en route. And on the subject of identification of navigation facilities, I wonder just how many of today's pilots know or can even decipher the Morse code identifiers?

Fuel management is, of course, a substantial portion of cockpit resource management. We all know that it is nothing less than pure stupidity to run out of fuel in an airplane, a situation for which there is no excuse whatever. You can't pull up to the nearest cloud and park while you walk to the closest service station for a can of gasoline, as you can in the case of an automobile. Even so, a surprising number of pilots find themselves in the position of having to make an off-airport landing with nothing but fumes in the tanks, or with an empty tank selected while plenty of fuel remained in another tank. One of the more sensible regulations the FAA has foisted on us is one that requires adequate reserves for both IFR and VFR day and night flying.

The whole concept of CRM is nothing more than dividing responsibility (who in the airplane or on the ground is responsible for doing what, when, and how), and maintaining the proper priorities. You know, aviate (fly the airplane—keep it upright), navigate (keep it pointed in the right direction), and communicate (talk last). That guy sitting on the ground in a room with no windows isn't going anywhere, and if your workload prohibits an immediate answer, so what? The guy operating the controls of the airplane is under the most pressure with the maximum workload at those times when he is required to be most precise and accurate: the takeoff and initial climb and the approach and landing phases of any given flight.

And on the subject of communications with a guy who isn't going anyplace, let me pass on a tip. I cannot listen to two people at the same time (my poor little brain would be overloaded). Therefore, I always pick up the ATIS (Automatic Terminal Information Service) on the hand-off—that is, either when Center hands me off to Approach, or when Approach hands me off to the Tower. My procedure is to tune the radio not in use to the next frequency (approach or local) and pick up the ATIS on the other radio. Then I switch over to the new frequency and check in, advising the controller that I have the ATIS. Although the controller is expecting me, he's not exactly holding his breath until he hears me check in, and nothing is happening so fast that I miss something important while I'm getting the ATIS between listening to the two facilities.

And on the subject of cockpit communications there is one paramount rule above all others—the meaning must be crystal clear to both parties to the communication. Do you recall the story of the tower controller who repeatedly ordered a student pilot who was doing touch-and-go landings to, "Go around. Disabled aircraft on the runway."? This instruction was repeated several times as the student continued her approach. Finally, she landed, taxied around the airplane on the runway, and took off. She had, indeed, gone around the disabled aircraft on the runway, just as instructed by the tower controller. Then there's the one about the pilot of a Boeing 707 who, on short final, decided to go around, so he shouted to the flight engineer, "Takeoff power!," meaning, of course, that he wanted max power for the go-around. The flight engineer took off power (that is he retarded the power levers to idle), and the airplane settled to the ground short of the runway. This particular 707 was crewed by a team of FAA employees.

In the "I Learned About Flying from That" feature in the October, 1996 issue of *FLYING* magazine, there is an excellent example of what can happen when there is a breakdown in CRM, this time in the failure to brief. Seems that the writer, a South Africa Air Force reservist had been away from the squadron for a while and required a currency instrument check which included recovery from a spin under the hood. While he was under the hood, the front-seat safety pilot and check airman in the North American AT-6 put the airplane in a spin and then said, "You've got her!" The pilot being checked kept waiting for the check pilot to shout, "Recover!," but he never did and the airplane kept rotating until it was dangerously close to the ground, at which time the check pilot grabbed the controls and initiated recovery procedures. Each thought the other was trying to play "chicken," but in fact while the pilot being checked was away, the procedure had been changed so that now instead of the check pilot telling the pilot being checked when to recover, the pilot being checked was supposed to automatically recover as soon as the controls were handed off to him. Nobody had bothered to inform him of this change, so he expected to be told when to recover. This lack of proper briefing could easily have resulted in disaster, and after the crash nobody would have been able to figure out why two experienced pilots spun in with no attempt to recover.

# VI Staying ahead

Bob Buck, in his book, *Flying Know-how*, says:

> *"Being ahead of the airplane." That's an expression we've all heard so much, we're apt to toss it off without paying attention to it. But it may be the most important expression in aviation, and certainly is worth time and talk. Being ahead is pretty obvious: It's a matter of planning in advance so nothing jumps up and surprises us. "Planning" and "being ahead" are synonymous. But what do we think when we plan? It's broken down into two parts:*
>
> *1.Advance*
>
> *2.What if*
>
> *"Advance" planning is simple. You are going to fly somewhere: How far is it, how much fuel will it take to get there, do I have the maps and equipment to do it?*
>
> *"What if" planning is different. It asks what if, when up there, the weather ahead goes sour, or the headwind is twice what expected, or an engine (maybe the engine) quits?*

*Actually you cannot do enough "what if" planning. But let's remember not to let "what if" planning turn us into nervous Nellies. We've got to look at "what if" with cool objectiveness, as something to prepare us for eventualities, and not something to worry us. It's one of the parts of flying that cannot be ruled by emotion because if "what if" thinking gets emotional, you'll never get off the ground—you'll be too scared! Neither will you ride in an automobile, or even get out of bed!*

Over the years, I have often been asked the question,"What is the most important, most valuable asset a pilot can have? What skill is the most useful?" I have invariably answered, "The ability to anticipate, to know what to expect next and to be prepared for it." We've all been told to make a plan and fly the plan. This does not mean we should be inflexible. I have observed pilots, prior to takeoff, plan a route and then insist on attempting to fly that route although it was literally impossible to do so. It is imperative that we strike a balance between anticipating what is coming next and being flexible enough to adjust our course of action as required by circumstances as they unfold.

Having gone through the decision-making process and reached a conclusion and determined a course of action, we must still avoid being so locked in to that specific mind-set that we lose the flexibility to change to a better course.

In a machine that is moving forward at two or more miles per minute, it is absolutely essential that our thinking be even faster. This is what is meant by staying ahead of the airplane. Once we permit it to get away from us and we find ourselves behind the airplane, we are in deep trouble. And it gets worse because as soon as we start to play catchup, we slip further and further behind until we find ourselves in a really hopeless situation.

Rod Machado advocates that instrument pilots in IMC (instrument meteorological conditions) be constantly thinking of the next two tasks. He says we should keep asking ourselves, "What are the next two things I will be doing?" I think this is an excellent idea, but I would take it one step further and apply this principle to all flights, VFR and IFR. If we are always attempting to anticipate what's coming up, we won't be caught by surprise, and I keep telling anybody who will listen, "There should be no surprises!" We must always be expecting the unexpected.

Thus, as we "walk" our communication radios (or set the next frequency into our standby on the "flip-flop"), we always have the next one ready. And as we walk our navigation receivers, we always have the next one ready, with the OBS (omni bearing selector) set to the next radial we will be following or to the localizer frequency we will be using. As I've said before, our airplanes are full of expensive gadgets, and I want all of 'em working for me. I want every needle in that airplane pointing to something useful. As I track along an airway, my number 2 navigation radio is tuned to a VOR off to the side. I center the needle (with a "FROM" indication), then move it to full-scale deflection ahead. When I have traveled 20 degrees along the airway, the needle will have traveled all the way over to the other side. I then rotate the OBS 20 more degrees and let it repeat the process. That way I know my approximate location on the airway, even without a DME, Loran C, or GPS.

And on the subject of the VOR receiver, are you all aware of the frequency range of the localizers? If you know that the odd tenths between 108 nothing and 111.9 are assigned to localizers, it will prevent you from becoming confused and tuning to the wrong facility, as I've seen several pilots do. (At PTK, the VOR is 111.0 and the LOC is 111.1, and on more than one occasion I've sat and watched students and applicants tune the VOR and fly right to the station when they thought they were following the localizer.) The VOR is 4.9 km northwest of the Airport.) This wouldn't have happened if they'd been aware of the frequency range of the localizers. I'm sure this sort of thing happens at many other locations as well, particularly those where the VOR associated with the airport and the LOC have somewhat similar frequencies and the VOR is located some distance from the airport.

# 6

# Responsibility

## I Who's in charge here?

One of the most difficult subjects an instructor ever has to teach is the awesome responsibility of being pilot in command of an airplane carrying passengers. 14 U.S.C. (FAR 1.1) says, "Pilot in Command' means the pilot responsible for the operation and safety of an aircraft during flight time." And according to Part 91.3 (a), "The pilot in command of an aircraft is directly responsible for, and is the final authority as to, the operation of that aircraft." This begs the question of the difference between acting as PIC and logging PIC time. Also, please note that "flight time" as defined in Part 1 means when you first board (or at least start up) an aircraft with the intention of flying it. This means that responsibility starts sooner than many people think. I've seen well-trained pilots bury their heads in the cockpit, shout, "Clear!" and turn the key to engage the starter without ever looking around outside to make sure the area is actually clear. Is this a responsible action? Of course not, but a great many pilots do it. The responsibility is the pilot's, but he or she is not acting responsibly when he or she does that.

In order to really know what it is to be PIC, it is first vital to understand that acting as pilot in command and logging PIC time may be quite different things indeed. And when this is coupled with the fact that each FAA region interprets the regulations as it sees fit (and, worse, many Flight Standards District Offices have their own interpretations), it's no wonder that confusion reigns. There can even be differences between the way individual inspectors in the same FSDO read and interpret the regulations.

At every pilot education clinic, flight instructor clinic, examiner clinic, flight instructor meeting, and examiner meeting, a substantial amount of discussion is devoted to the logging of flight time, yet the vast majority of attendees come away even more confused than when they arrived. And these FAA Flight Standards Aviation Safety Inspectors who conduct these meetings are the guys who are looked to as experts. Here's an example: after a half-day discussion of PIC rules at a recent CFI revalidation clinic, a controversy arose regarding a specific rule dealing with the logging of PIC time. During a break the speaker called the FAA's Flight Standards office in Washington to get a definitive ruling. Meanwhile a call was placed to the local FSDO with the same question. You guessed it—two diametrically opposite answers were received, both from FAA authority. What this tells us is that we'd better be in compliance with whatever local jurisdiction we happen to be in, or at least be in a position to defend ourselves based on some higher authority. If it is possible to interpret a regulation in more than one way (and it always is), you can be sure that it will be done.

Much of the confusion about logging time, or other FARs for that matter, results from the fact that many of the rules grew up as a patchwork. First there is an event causing alarm within the FAA. Next comes a knee-jerk reaction to the event, along with an immediate call for a regulatory Band-Aid to cover up the problem—and the aviation community is stuck with another regulation that may well cause more problems than it cures, all in the name of safety.

Not all PIC flight-time-logging rules are confusing, though they may be illogical. For example: prior to the private pilot checkride, a student pilot may not log any time as PIC, even the hours when they are alone in the airplane. That time can only be logged as "solo." See what I mean? The rule is clear, but logic is lacking. Just who is in command when the student flies solo? It may very well depend on what a court has to say, in spite of the fact that the FAA says the student was solo but not PIC. This is in direct conflict with the definition of PIC above. Obviously somebody is in command when a student is out solo, but the FARs don't tell us who it is. In many court and NTSB administrative law cases resulting from an accident or incident that occurred while a student pilot was solo, the student's instructor was held to be in command (or at least responsible) even while sitting on the ground; in others, the student, even though not PIC by FAA definition, was found to be legally responsible for the operation of the airplane, and thus PIC.

Instructors log PIC time when flying with a student but can't log as PIC that time when their students fly solo, even though the CFI may be legally in command of the flight. Does that make sense? It does when you think in terms of just who is responsible. Instructors are often charged with responsibility for their students' actions.

Once individuals hold recreational or higher pilot certificates, the FARs change to allow individuals to log as PIC that time they're manipulating the controls while receiving instruction. The regulations permit pilots who are qualified to fly the airplane to log as PIC that time during which they are the "sole manipulator" of the controls while receiving dual instruction. In that case, the advanced student (beyond the Student Pilot Certificate) logs time as both dual and PIC. Thus we see a situation in which when a student pilot is solo, there is no PIC in the cockpit, and when an advanced student is receiving instruction there are two PICs in the cockpit, or at least two pilots logging PIC time. This apparently contradictory FAA attitude came about when the requirements for the commercial certificate were expanded to include "complex airplane" experience. No FBO could get insurance coverage for a complex airplane to be flown solo (or as PIC with passengers) unless the renter pilot—frequently an advanced student—had far more time in type than the minimum the FAA requires for the commercial certificate. Therefore, the FAA changed its interpretation of the rules to permit the pilot to log both PIC and dual in that situation so that when advanced students completed their commercial training, they would have enough PIC time in retractables (and other requirements) to be insurable. This interpretation opened the door to further expanding the difference between acting as PIC and logging PIC.

Logging PIC while flying IFR can be complicated. Even though FAR 61.51(c)(2)(i) permits pilots to log as PIC the time they are the sole manipulator of the controls of an airplane for which they are rated (qualified to fly by virtue of the certificate they hold), FAR 91.173 states that, "No person may operate an aircraft in controlled airspace under IFR unless that person has (a) filed an IFR flight plan; and (b) received an appropriate ATC clearance." (Note that the regulation refers only to "controlled airspace.")

Obviously, instrument-rated pilots may fly in IMC in uncontrolled airspace without a flight plan, but how about noninstrument-rated pilots? (I can't find anything in the regs to prohibit them from flying in IMC in uncontrolled airspace!) To fly an aircraft on an IFR clearance, the PIC must have a current and valid instrument rating. This PIC rules conundrum has consistently been interpreted to mean that no pilot may log as PIC any time in which the flight is in IMC (instrument meteorological conditions; that is in cloud) or on an IFR ATC clearance unless that pilot holds a current and valid instrument rating. An instrument rating is only valid when the pilot has met the recent experience requirement. Thus, if your IFR currency has lapsed and you fly on an IFR clearance with an IFR current pilot in the right seat, the person in the right seat has to be PIC, and you can't log it as such even if you are the "sole manipulator." If, however, you fly simulated IFR without an actual IFR clearance, you can log both PIC and simulated IFR time toward currency, and what the safety pilot logs is open to question. The only time that a noninstrument-rated or noncurrent IFR pilot may log PIC when flying in IMC or on an actual IFR clearance is when dual instruction is being administered by an appropriately rated flight instructor. And in that case both pilots log PIC and actual IFR for the flight time in cloud. Obviously, the CFI can't log simulated IFR time, even though he or she is PIC on a flight on an IFR clearance in VMC where the pilot flying is wearing a view-limiting device. So when a CFII and rated pilot fly on a clearance in visual conditions we have two PICs but only one pilot logging simulated instrument time. Pop into the clouds and we have two pilots logging actual IFR time.

For a pilot to log SIC (second-in-command) time, either the operation (such as FAR Part 135 passenger carrying under IFR without autopilot authorization) or the aircraft (by virtue of its type certificate) must require a copilot or second in command. Some FSDOs hold that safety pilots flying with an instrument student or instrument-rated pilot maintaining currency should log their time as SIC, based on the fact that the operation requires their presence. Others maintain that both the acting PIC and the safety pilot should log their time as PIC. And still others say the safety pilot can do it either way. How's that for inconsistency? Again, it all depends on whose interpretation you choose to accept.

Logging time is essential to advancing one's pilot ratings, retaining currency, and in general by establishing one's level of pilot experience (perhaps to meet an insurance requirement), so hours recorded in a book are vital to all pilots. But the most perplexing and crucial questions of command authority arise, of course, when there are two or more pilots aboard an aircraft involved in an incident, accident, or violation. When there is an accident and lawsuits begin flying instead of airplanes, we enter a whole new ballpark to play a different game—the court game.

Because aviation by its very nature frequently involves the crossing of state lines, the federal government has preempted the law regarding all things aeronautical. Thus, if a state aviation commission should make rules that conflict with the FARs, the FARs take precedence. However, in the case of a civil suit brought in federal court, the federal court will apply the law of the jurisdiction (state) where the suit is brought. It may not seem fair to change the rules for a civil trial, but we see the differences between criminal and civil legal actions all the time. We all, for example, have constitutional protection against self-incrimination in a criminal trial, but no such protection in a civil or administrative case (see Chapter 13). The federal court's deference to jurisdictional procedures in civil suits opens up a whole new bunch of interpretations regarding who was PIC at the time of the accident, and if all aboard were killed in the crash, the issue can become very important indeed. It also can, and usually does, become extremely complex and difficult to sort out. I heard a judge once remark, "It will take at least two years to sort out the players."

The law in some states holds what lawyers call a "rebuttable presumption" that the occupant of the left front seat was the pilot flying at the time of the crash. This means that the court starts out assuming that the guy in the left seat was manipulating the controls when the accident occurred and was acting as pilot in command, but this assumption can be overturned by evidence to the contrary. In some states the law says that the holder of the highest grade of pilot certificate was in charge at the time of the accident, no matter where that pilot was sitting. And in some jurisdictions, this is again a rebuttable presumption, while in others the court accepts this as an absolute fact. In many cases, if there is a CFI aboard, the CFI is held responsible, but this too may, in some jurisdictions, be rebutted by evidence to the contrary.

Since the federal government wants to avoid responsibility for any airplane accident, and FAA inspectors are employees of Uncle Sam and designated pilot examiners are representatives of the administrator when acting in their capacity of examiner, there is a special FAR governing the status of inspectors and examiners while conducting flight tests. FAR 61.47 says inspectors and examiners are specifically not PIC in the flight test situation. Their status is that of "passenger/observer." Even so, in spite of this very specific FAR, some courts have held examiners responsible for accidents that occurred during the course of a flight test, but so far as I know, never an inspector who is an actual government employee, a prime example of the government protecting itself. Other courts have refused to hold examiners responsible based on FAR 61.47. How confusing can it get?

Incidentally, student pilots whose certificate specifically states "Passenger Carrying Prohibited" both act as and log as PIC the time flown during their flight tests, and they have a passenger aboard (the inspector or examiner whose status is passenger/observer). The theory under which student pilots may not log PIC, but only "solo" when they are alone in the airplane is based on FAR 61.51, which states that a recreational, private, or commercial pilot may log PIC time when he or she is the sole manipulator of the controls of an aircraft for which he or she is rated (holds a certificate for that category and class), and a student pilot does not have such a rating. Who knows just why that theory goes out the window when an examiner gets in the right seat to administer a flight test as an official "passenger/observer?"

At least one jurisdiction (state), by statute, holds the person named on the flight plan responsible as PIC in case of an accident or violation. Then, of course there's the legal concept of "vicarious liability" applied by some states which makes the owner (or owner/operator) liable for an accident even though the owner may not be present during the occurrence. If the owner/operator happens to be aboard at the time of an accident he or she may also be deemed to be PIC even if he or she wasn't at the controls or even a rated pilot. Perhaps a court may find he or she exercised "command authority" by giving orders to the person in the left seat who was driving the airplane. Though such a concept may seem laughable to pilots, the captain of a ship does not operate the controls but is certainly held responsible for the safety of the vessel. Because command authority and control manipulation are not always the same, the term "pilot flying" and "pilot not flying" are now commonly used in cockpit crew training instead of the terms "captain" and "copilot." The captain is always in command but not always the pilot flying. Of course, this is merely the use of semantics to cover the issue.

When a case involving multiple pilots aboard gets to the trial stage, the court may look at such questions as: (1) Who had access to the controls? (2) Who sat in the left front seat—the traditional location of the PIC? (3) Who held the highest grade of pilot certificate, or who had the most overall experience, or who had the most experience in make and model, or in the existing flight conditions? And according to Barbara J. Gazeley, writing in the *Lawyer Pilots Bar Association Journal*, "Who was the pilot at the time of takeoff, and how long after takeoff did the accident occur?" In some cases, she points out, "the court has found that it would be unlikely for a more qualified pilot to allow a novice or passenger to operate the controls during the takeoff and landing phases of flight." This implies that command may change after the flight gets underway and levels off in cruise.

Accident investigators will look at the physical evidence (position of controls and so forth) in an effort to determine just who was flying at the time of the accident. (The hands on which body are clutched on the yoke?) What was the weather? If it was IMC, who among those on board was instrument rated? Which of the pilots was working the communications radios? (Some courts have held the opinion that the nonflying pilot would most likely be the person to talk on the radio, while others have taken the opposite position.) Any or all of these factors may be considered by a court in its determination of the question of pilot in command, and thus ultimate responsibility for the flight. In at least one case, a pilot who had permitted his medical to lapse delayed a flight until another pilot with a current medical could come along to "legitimize" the flight. Even so, the court held the unqualified (no current medical) pilot to be responsible for the accident or violation. On the other hand, an owner/operator (in one case a corporation) who unknowingly hired an unqualified pilot (no pilot certificate whatever) has been held liable for his actions. Nobody ever said life is fair. Can you imagine anyone hiring a "pilot" without knowing that he or she held a certificate?

Thus it may not matter at all who logs PIC time, or even who was actually flying the airplane, in determining responsibility for the flight, even though the FAA says the PIC is responsible. A court may have something different to say, and in the final analysis, what the court says is what really matters.

Once again, a careful reading of FAR 61.51(c)(2)(i) indicates that a pilot who is qualified to fly the airplane by virtue of a certificate and rating (higher than student certificate and rated for that category and class) may log as PIC that time during which he or she is the sole manipulator of the controls, provided he or she is also qualified (authorized) to fly by "the regulations under which the flight is conducted." To determine just what regulation applies we must look further—if under FAR Part 91, is the flight VFR or IFR? Is it a commercial operation, or otherwise "for hire?" (The FAA's definition [interpretation of the term] "for hire" is another weird one, but not a subject for this discussion. See Chapter 15.) In applying the FARs, we must frequently consult the rule we think applies, and then cross-reference it to another to determine just how it applies.

The bottom line is this: if there is any opportunity for the regulations to be interpreted in more than one way, we can be sure they will be, and in some cases in two diametrically opposed ways. If you find all this confusing, rest assured that you're not alone.

As this is being written, there is a huge NPRM (Notice of Proposed Rulemaking) published, which, if some of the proposed changes go into effect, will drastically change how pilots log PIC time. Among many other changes proposed is one which would limit to a single person the right to log PIC time. As it now stands, there can be as many as four pilots on a single flight simultaneously logging PIC (one of 'em does not even need to be a pilot).

1. An advanced student in the left front seat manipulating the controls.

2. A CFI in the right front seat training the one in the left front seat.

3. An ATP in one of the rear seats.

4. The "owner/operator" in the other rear seat.

And all of 'em logging PIC, even though only one is "acting as PIC." How do you like them apples?

## II Just how far will they reach?

The question of which of two pilots is pilot in command at any given time is determined by what the person or persons in authority say and just how he, she, or they interpret their own rules with respect to who is PIC. And different authority will frequently occupy different positions on this matter. The following case is an excellent example serving to illustrate this principle:

It is well established that student pilots will sometimes fly with certificated pilots, friends perhaps, who are not certified flight instructors, and on occasion student pilots will even manipulate the controls of the airplane. The students thus gain experience, but they may not log the time since they are neither alone in the airplane nor are they being given dual instruction by a properly certified and rated flight instructor.

On July 20, 1980, Eugene R., a student owner flew his Cessna 150 from Lancaster, California to Bakersfield, California. He occupied the left seat and Dave K., a certificated private pilot was in the right seat. They had flown together on several previous occasions, and student pilot Eugene R. had never logged these flights. Dave K. had always logged these flights as PIC, even though the student was manipulating the controls. On the July 20 flight, with Dave reading the (wrong) frequencies to Gene, who was controlling the airplane, they penetrated the Bakersfield Airport Traffic Area and landed without making contact with the Airport Traffic Control Tower. They also cut off a Cessna 441, which was on a 1-mile final and which was required to take evasive action to avoid a collision. (It executed a go-around.)

When it became evident that a violation might be issued, private pilot Dave decided he was a passenger, and that student pilot Eugene was pilot in command. (Note: Some friend, huh?) Believe it or not, the local Flight Standards District Office (the friendly feds—the FAA) bought this nonsense and sought an emergency revocation of Eugene R.'s Student Pilot Certificate. (This was an emergency?) On November 20, 1991 the then FAA administrator issued an Emergency Order of Revocation in the case of student pilot Eugene R., citing the following allegations:

1. You are now and at all times mentioned herein were the holder of Student Pilot Certificate No. BB-006198196.

2. On or about July 20, 1990 you were the pilot in command of Civil Aircraft N5790E, a Cessna 150, operating on a flight from Lancaster, California to Bakersfield, California.

3. Incident to said flight you flew N5790E from Bakersfield Skypark to Meadows Field, a controlled airport.

4. You were not authorized by your instructor to fly cross country on this date, and you had not received authorization to land at Meadows Field.

5. You entered the Meadows Field Airport Traffic Area without contacting the Airport Traffic Control Tower and without receiving prior clearance to do so.

**6.** During your initial approach to land at Meadows Field, you crossed in front of Civil Aircraft N1208A, a Cessna Model 441, which was on a 1-mile final, and forced that aircraft's pilot to maneuver to avoid a collision with you.

**7.** Incident to said flight, you operated N5790E with a passenger aboard.

**8.** Incident to said flight, you failed to have your pilot certificate, your current airman medical certificate, and logbook in your personal possession.

**9.** You failed to produce your current pilot certificate, your current airman medical certificate, and your logbook when they were requested by the airport security officer.

**10.** Your operation of said aircraft was careless or reckless so as to endanger the life or property of another. (Note: They almost always throw this one in, often to use as a bargaining chip in compromising a charge or penalty.)

By reason of the foregoing circumstances, you:

**a.** violated Section 61.3(a), in that no person may act as pilot in command of a civil aircraft unless he has in his personal possession a current pilot certificate.

**b.** violated Section 61.3(c), in that no person may act as pilot in command of an aircraft under a certificate issued to him unless he has in his personal possession an appropriate, current medical certificate.

**c.** violated Section 61.3(h), in that each person who holds a pilot certificate, medical certificate shall present it for inspection upon request of a state or local law enforcement officer.

**d.** violated Section 61.89(a)(1), in that a student pilot may not act as pilot in command of an aircraft that is carrying a passenger.

**e.** violated Section 61.93(d)(2)(1), in that no student pilot may operate an aircraft in solo cross-country flight unless an instructor has endorsed the student's logbook and without obtaining the instructor's review of the student's preflight planning and preparation, attesting that the student is prepared to make the flight safely under the known circumstances and subject to any conditions noted in the logbook.

**f.** violated Section 91.129(b), in that no person may, within an Airport Traffic Area, to, from, or on an airport having a control tower without maintaining two-way radio communication between the aircraft and the control tower.

**g.** violated Section 61.51(d), in that a pilot must present his logbook for inspection upon reasonable request by a state or local law enforcement officer.

**h.** violated Section 61.51(d)(2) in that a student pilot must carry his logbook with him on all solo cross-country flights as evidence of the required instructor clearance and endorsements.

**i.** failed to exercise the degree of care, judgment, and responsibility required of the holder of a Student Pilot Certificate.

**j.** have demonstrated that you presently lack the qualifications required of the holder of a Student Pilot Certificate.

On December 17, 1991, an emergency revocation hearing was held at the NTSB office before an administrative law judge (ALJ). The only real issue before this tribunal was whether or not student pilot Eugene R. was in fact pilot in command on the flight in question.

The government (FAA) contended that since Eugene R. occupied the left seat and was the sole manipulator of the controls, he was indeed pilot in command and thus responsible for the entire flight since FAR 91.3(a) states, "The pilot in command of an aircraft is directly responsible for, and is the final authority as to, the operation of that aircraft."

Counsel for the Respondent (Eugene R.'s lawyer) maintained that because his client had on several previous occasions flown with private pilot David K., had occupied the left seat and manipulated the controls, and on each occasion they had mutually agreed (either explicitly or implied) that Dave was PIC and not Gene, and in each such case Dave had logged the time and Gene had not, it would be unreasonable to determine that on this occasion Gene was PIC. He also contended that to find otherwise would be detrimental to aviation in that it would prevent student pilots from gaining valuable experience by flying with certificated pilots and operating the controls, even though they may not log the time.

On hearing the evidence and testimony of the parties and witnesses, the ALJ found in favor of the FAA and upheld the order of revocation. The judge determined that the FAA had met the burden of proving that student pilot Eugene R. was acting as pilot in command. Apparently the court never addressed the question as to whether or not an airport security officer is a "state or local law enforcement officer" and thus authorized to demand the right to inspect Eugene's certificates.

Gene's attorney simply could not believe this decision, and he took the matter up on appeal to the full board (NTSB). In fact, he found it incomprehensible that the FAA had even claimed that the student was PIC in the first place. It was his contention that the violation should never have been issued against Eugene R.

The issues on appeal were: one, Did the FAA sustain its burden of proving by a preponderance of reliable, probative, and substantial evidence that Eugene R. was in fact pilot in command at the time of the alleged incident? and two, Was revocation of a Student Pilot Certificate under these circumstances an excessive and unwarranted punishment?

Eugene's lawyer submitted to the NTSB that the decision of the administrative law judge upholding the revocation of Eugene R.'s Student Pilot Certificate should be overturned for the following reasons: one, The FAA failed to sustain its burden of proving Mr. R. was in fact pilot in command at the time of the alleged incident; two, The ALJ disregarded the evidence, which clearly proved the certificated private pilot, David K. was in fact pilot in command because of his actual and implied assumption of the responsibility for the operation and safety of the flight in question; three, The ALJ disregarded competent expert testimony from a professional flight instructor and FAA accident prevention counselor, who asserted that Mr. R. did not violate any laws and was not pilot in command of any of his numerous flights with Mr. K.; four, The ALJ disregarded case authority which he himself cited as precedent at the hearing in favor of cases which are not on point and do not apply; five, Revocation of a student certificate under these circumstances is an excessive and unwarranted punishment; and six, Allowing the ruling of the ALJ to stand would have a deleterious effect on the ability of student pilots to gain experience during the training process.

Student pilot R.'s attorney pointed out to the board that the following facts had been established at the December 17, 1991 hearing before the ALJ: one, private pilot David K. had flown with student pilot Eugene R. on several occasions, on which flights he conceded to being pilot in command, despite the student being the sole manipulator of the controls; two, He knew Mr. R. was a student pilot; three, He is aware that a student pilot cannot fly with passengers aboard; and four, He admitted that on the flight in question he would automatically take control of the aircraft in the event of an emergency (Note: this fact alone should have been enough to make a determination that Mr. K. was PIC). The attorney argued that all of this, along with case precedent and testimony at the hearing, established that Eugene R. was not pilot in command and thus could not be responsible for the alleged violations.

At the original hearing, the ALJ himself cited the 1975 case of *Administrator v Dye* in which the board held "if a student pilot is accompanied by another individual who holds a higher ranking pilot certificate, then regardless of who is at the controls, the other pilot will, under most circumstances, be considered pilot in command." Also cited was the 1982 case of *Administrator v Fields*, where the board held that the right-seat pilot, acting as safety pilot, was pilot in command, despite the left-seat occupant having manipulated the controls, handled the radios, etc. In that case the board concluded "the fact that a pilot is sitting in the left seat is not necessarily determinative of which of the two pilots is pilot in command. Rather, our conclusion that Respondent was pilot in command rests on his overall responsibility for and control of the flight in light of the circumstances."

However, the administrative law judge chose to ignore the Board's rulings in these cases, instead relying on two cases cited by the FAA, neither of which really applies, both 1983 cases, *Administrator v McCartney* and *Administrator v Fleischman*. In McCartney the Respondent was found to be pilot in command even though he was operating with a suspended Commercial Pilot Certificate and his companion had a valid certificate because he was the person "with overall responsibility for and control of a flight." This case differs from the present matter in that the Board was not dealing with a student pilot, and the properly certificated pilot who was aboard did not know that McCartney's certificate was suspended. The judge ruled in the Fleischman case for the holding that an aircraft owner has the burden of proving that he is not pilot in command, there being an inference (rebuttable presumption) that he is pilot in command merely by virtue of his status as owner. This may be a reasonable inference when the owner is a certificated pilot, and

there is evidence that nobody else on the airplane is known to hold
a pilot certificate. (These were the facts in Fleischman and, again,
they do not apply to the instant case.)

With respect to the testimony, evidence presented at the December
17, 1971 hearing before the ALJ also proves that private pilot Dave K.
and not student pilot Eugene R. was pilot in command. First,
Mr. K. testified that he knew that Mr. R. was only a student pilot and
he also knew that student pilots are prohibited from carrying passengers. He stated that he had been a private pilot since 1968, and
that during each of his flights with Mr. R. he had in his personal
possession a pilot certificate, a valid medical, an FCC permit, and a
current Biennial Flight Review as required by the regulations. He
testified that he had approximately 3,000 hours total time flying experience, 500 in Cessna 150s, and he had ready access to the dual
controls in Mr. R.'s Cessna 150. His only basis for claiming that
Mr. R. was pilot in command was because Mr. R. occupied the left
seat. He further testified that he had flown with Mr. R. on several
previous occasions when Mr. R. occupied the left seat, and yet he
considered himself to be pilot in command. Finally, his state of
mind was such that in the event of an emergency he would automatically take over the controls.

Eugene R.'s flight instructor testified that he was aware that Mr. R.
and private pilot Dave K. flew together and that he knew of no regulation prohibiting a student flying in the left seat of his 150 with a
private pilot in the right seat. Mr. Eugene R., the student-owner, testified in his own defense that he had flown with David K. at least a
dozen times, usually occupying the left seat himself, and that, based
on past experience, he assumed that Mr. K. would be PIC. He also
testified that at the time the aircraft flew into the Meadows Field Airport Traffic Area, it was Mr. K. who was providing the improper frequencies to him, that he did not log any time when he was flying
with another certificated pilot such as Mr. K., and that if any problems arose he would give over the controls to Mr. K.

Mr. R. called as an expert witness a certain Ken K., an ATP pilot with
a CFI certificate with ratings for Airplanes and Instruments, and a
Ground Instructor Certificate as well. Ken K. is a professional flight
instructor and an FAA-designated accident prevention counselor
who has administered in excess of 5,000 hours of dual instruction.
Ken K. rendered his expert opinion based on experience, FARs, and
a poll of several designated pilot examiners. He opined that Mr. Eu-

gene R. was not PIC, since a private pilot who knowingly steps into the aircraft with a student pilot assumes command.

At the December 17, 1991 hearing both the FAA and the ALJ made very clear the fact that, all else aside, the revocation was sought by the FAA and granted by the ALJ because Eugene R. was PIC of an aircraft carrying a passenger although he held only a Student Pilot Certificate and was thus prohibited from doing so.

On appeal, Mr. R.'s attorney also argued that revocation was excessive punishment (in legalese, too great a sanction) for the inadvertent offense (if, indeed, any offense should be found to exist on the part of Eugene R.). He also argued that common sense dictates (Note: When did common sense have any relationship to anything the FAA ever did?) that the private pilot was PIC, and that to find otherwise would have a "deleterious effect on the ability of student pilots to gain experience during the training process. If this ruling is permitted to stand a student pilot will not be able to get in an airplane with a licenses (sic) pilot and touch the controls for fear that his license (sic) will be revoked for flying as PIC with passengers."

On January 19, 1992 a unanimous National Transportation Safety Board overturned the ruling of the administrative law judge and restored Mr. Eugene R.'s Student Pilot Certificate to him. Common sense and justice finally prevailed! But one cannot help but wonder how on earth the FAA could have brought this action in the first place, and even worse, how the ALJ could have found in favor of the FAA and against Mr. Eugene R. It is literally beyond comprehension how this matter was permitted to go as far as it did, forcing Mr. R. to defend himself against these ridiculous charges. However, it does provide us with a perfect example of the concept that PIC is what the appropriate authority (in this case the full Board) says it is.

Most of this material was taken from an article by Charles M. Finkel, Esq., Mr. Eugene R.'s lawyer, who so ably defended him. The article appeared in the *Lawyer Pilots Bar Association Journal*.

# III Responsibility under the law

Anybody can sue anybody else at any time for any reason, and in America today anybody is likely to do just that. The popular thing to do these days is to file a lawsuit, and although most of the blame for this falls on greedy plaintiff's lawyers, the public at large seems to have caught on to this easy way to acquire money without working for it. No matter how baseless or frivolous it may be, if somebody thinks he or she has even a remote chance of collecting something, he/she will file a lawsuit. And they'll have no trouble finding an attorney to undertake it for them. The really bad thing about it is that when the word "airplane" is mentioned to a jury, they just start handing out money in vast quantities. I guess this is because in the collective mind of the uninformed public, airplanes are inherently dangerous instrumentalities, plus the fact that generally people who don't know better hold the belief that those of us who own airplanes are extremely wealthy and can easily afford to lay out huge sums to settle these suits, rather than be put to the burden of defending ourselves.

How bad has it gotten? It has reached the point where we have to go far out of our way to protect ourselves against the possibility of being sued. Not too many years ago, when student pilots would reach that point in their training that the instructor believed the student was ready for solo, the instructor would merely make an entry in the student's logbook stating, "Okay for solo." Now the instructor carefully documents every single presolo maneuver and then uses the words, "(student's name), checked and found competent for solo in the pattern at (airport) in (make and model of airplane) on (date)." This, only after having documented the fact that the student has satisfactorily completed all required presolo work. And it can be just as bad for the pilot examiner.

I am familiar with a specific case in which an applicant came to an examiner for a private pilot practical test. The applicant came from another airport some 30-odd miles away. The checkride was satisfactory and the examiner issued the applicant his certificate. A year later, the (now) private pilot went to the operator at the airport where he was trained, and from whom he had been renting airplanes, to rent an airplane for a trip. He had three heavyweights with him, and they loaded a PA28-140 (poor little Cherokee 140) with a lot of heavy hunting equipment and the four of them started to board the airplane.

Seeing this, the operator rushed out of his office and shouted, "Stop! Do a weight and balance!" Whereupon the pilot and his passengers off loaded the airplane, and the heaviest of the four put all their equipment in his car and left, saying he'd meet the rest of them at the destination. The other three got back into the airplane and took off. You guessed it. They flew about 20 miles toward the destination and landed at another airport where they reloaded the equipment and the fourth fat guy. On the takeoff attempt, they crashed, killing all four, a not-unexpected result (see Chapter 1).

The next thing that happened was the pilot examiner received a summons. He had seen the pilot only once, well over a year before, and he was being sued by the estate of the pilot and those of the passengers for having issued a pilot certificate to the dumb guy who had killed himself and his friends! In this case, the defendant examiner was fortunate in that he worked for an FBO who had insurance which provided the examiner with a defense attorney. Even so, it cost the pilot examiner some 70-odd thousands of dollars out of pocket in legal fees before he was dismissed out of the lawsuit. (He retained his own counsel to defend him in case there should be an excess award over and above the policy limits of the insurance on him carried by the FBO.)

In order to get himself dismissed out of the suit, the examiner took deposition testimony from some 80 pilots that he had examined over the years, all of whom testified that as part of their practical tests they were required to work a weight-and-balance problem.

I have seen other lawsuits just as far fetched. How about this one: a brand new private pilot loaded his family (six souls including two very small children) in a four-seat Cessna 172 and flew some 300 miles to visit relatives. He landed too fast and too far down the runway at the destination, a single strip in the mountains of West Virginia. After leaving almost 60 feet of rubber skid marks on the runway, the pilot decided to attempt a go-around. He applied power and tried to take off with full flaps extended and carb heat on. The airplane staggered into the air, got about 50 feet off the ground and a quarter mile off the departure end of the runway when it stalled and crashed in a ravine, killing all aboard.

The estate of the family filed suit against the FBO that had rented him the airplane, claiming among many other things, that he had failed to provide parachutes for the pilot and passengers, and had he done so, they might have survived! The plaintiffs' attorney totally ignored the fact that there were six people in a four-place airplane and all the other things the pilot did wrong.

I have lots more such stories, but I'm sure you have the idea by now. Although the examples cited above are cases directed against an individual and an FBO, the most common victim of these frivolous lawsuits are the airplane and component manufacturers. After all, they're the ones with money, and the plaintiffs are going to search out the "deep pockets" to target with their lawsuits. The cases against the manufacturers are based on such flimsy claims as, although the airplane manual and placards warn the pilot not to mismanage his or her fuel, they don't say you might die if you do mismanage your fuel.

I'm familiar with a case in which a pilot knowingly flew into known heavy icing conditions in an airplane with no ice protection whatsoever. When the airplane crashed, both occupants were killed. The estate of the pilot then sued the manufacturer of the fuel injection system, the engine manufacturer for selecting that fuel injection system, and the airplane manufacturer for selecting that engine with that system, claiming that what brought the airplane down was carburetor ice, this in an airplane that didn't have a carburetor! In another case, a pilot ran out of gas and crashed, killing himself and his passenger. The passenger's estate filed suit against the airplane manufacturer, claiming the fuel system of the airplane was poorly designed.

With this kind of activity going on, is it any wonder that we in aviation have to do all kinds of things to protect ourselves against potential legal action? Of course, this is not just true of aviation, but aviation is the most extreme example with which I am familiar.

# 7

# Other good stuff

## I How to buy a used airplane

Since in the last 40 years or so I have bought a substantial number of airplanes (more than 40 of them) myself and been involved in helping others in the purchase of a great many more aircraft, I have often been asked to share my experience in the acquisition of airplanes. The best advice I can offer regarding the purchase of an airplane is: don't do it! Unless you are going to be flying 150 hours per year or more, unless you have absolutely unlimited funds, or unless you are willing to indulge yourself by wasting money on an expensive toy, you are much better off renting. There is just no way to delude yourself into believing that the private ownership of an airplane makes economic sense. I'm not talking here about the individual who is buying an airplane for business use, but rather the occasional pleasure flyer.

However, if you choose to disregard the above advice and are determined to go ahead with the foolish notion that aircraft ownership is for you, there are a number of important steps to take to ensure that you will be happy or at least satisfied with your purchase. The most important point to keep in mind is: take your time. Don't get in a hurry. As a general rule, airplanes are easy to buy and hard to sell. Keep this fact in mind at all times. Starting from this point on, I'm addressing any airplane buyer, business and pleasure.

The first step, once you have convinced yourself that this is something you really want to do, is to define the mission. Ask yourself just what it is that the airplane you want to own must be capable of doing for you. How many seats do you require? If you usually fly alone, or with only one other person, a two-or four-place airplane is likely to be adequate. Not only is the initial investment in the aircraft much greater with the addition of seating capacity, but the cost of insurance in phenomenally higher with each added seat.

You must next ask yourself how fast you really need to go. Over 120 or so knots of speed, and the price skyrockets. The next break seems to be at about 150 knots. And then there's another leap in price at the 170-knot range.

Do you require the dependability of being able to fly IFR? If so, you will look at only IFR-equipped airplanes, or have enough cash in reserve to properly equip the airplane you buy. You can buy a run-out old airplane with minimal equipment and spend a fortune cleaning it up and equipping it the way you want it, or you can find one in the shape and with the equipment you want and spend your fortune on it. Do you get what I'm saying? Either way you must be prepared to invest a huge sum of money.

Twin or single? This is a question individual buyers must answer for themselves. We all know the statistics. More people die in twins than in singles proportional to the amount of exposure. When I was trained, every landing was a dead-stick landing. This is because training is running 10 or more years behind the state of the art, and 10 years prior to the time I was trained, you could toss a coin every time you took off as to whether or not you'd have an engine when you came back. Today this factor is not that much of a concern. The engines are infinitely more dependable. So your choice might very well be a single unless you need so-called all-weather capability, and unless the redundancy of the second engine is so appealing that it outweighs the tremendous additional cost.

Once these questions are all answered, you have defined just what your new toy must do for you. The next step is deciding which among the various aircraft that meet that specific mission profile is for you. I am frequently asked which I prefer, a high-wing airplane, or a low wing. My answer has always been, "I've flown them with the wing on the top, on the bottom, on both the top and bottom, and even some with a wing in the middle, and they all do the job for which they were designed. The only question I need to answer is, "Which specific airplane will best serve my purpose today?". In other words, what is the purpose of the flight I'm about to embark on right now, and which airplane will do this job the best?" Not having a fleet of airplanes available to you, you can't do that. You must make a compromise. Decide what you will usually be doing, and select an airplane make and model from among those that fit your personal use as defined above.

Now that you've come this far, it is time to start shopping. You know the make and model you're looking for and you know your price range. And on the subject of price, you know just how much you've budgeted for this project. Do not permit yourself to be swayed into exceeding that amount! Start studying the ads. If you know a local dealer or broker, enlist his or her help in finding the airplane for you. It may not be fair to the honorable new person starting out, but you should try to deal with a dealer or broker who has been around for a while, one who has had repeat business, who has sold a series of airplanes to the same buyers. They wouldn't come back to that dealer if he or she hadn't treated them fairly on previous deals. If you are going through a dealer, talk with some of his or her previous customers. Are they satisfied, even happy, with the deal they got?

Now for a few specific tips. If you're shopping on your own and you travel to a distant location to look at an airplane you've seen advertised, do not be afraid to turn around and go home if you find it is not as advertised. Many airplane purchasers have regretted for years the fact that they pressured themselves into buying the wrong thing just because they had gone a long distance from home to see an advertised airplane that turned out to be not quite as advertised.

All airplanes have a history, and this can be traced through the maintenance records (aircraft logs, engine logs, prop logs, FAA 337 Forms — Major Alteration and Repair). Even these records are no guarantee of the condition of the airplane. Much as I hate to inform you of this, the fact is that it is not altogether unknown for work to be done without being recorded, and for records to be falsified! Get a reliable mechanic to go through the records with you! Use the same person who is doing the prepurchase inspection for you. The very best money you can spend is what you pay the mechanic to do the prebuy. And don't use the seller's own mechanic.

If the records show that the airplane has had major damage history, it should be reflected in the price. Depending on how far back in time the damage occurred, the airplane will be priced considerably less than a comparable make and model with no damage history. This may make it a real bargain. If it has been properly repaired, it may very well be better than it was when it was built by the manufacturer. If this is the case, and it is priced well below the market, and you intend to keep it for a long time, it may be a very good buy. You'll get the benefit of the depreciated value because of the prior damage, but by the time you're ready to sell it and upgrade, the damage will probably be so far in the distant past that the current value won't be affected, or will be only slightly affected. Note that I said "properly repaired." I have twice bought airplanes that had damage history and the repairs signed off by certificated mechanics (one a reputable repair station) which had not been properly repaired. In both cases the airplanes required extensive (and expensive) reconstruction before the FAA would consider them airworthy. This was true even though one of those two airplanes had gone through no less than eight annual inspections after the bad work was done. Of course, the annuals were signed off by the same shop that had done the improper work. And the guy who did the prebuy inspection for me (an excellent mechanic with IA) made the erroneous assumption that since a reputable certified repair station had made the repairs, it had been done properly. It was only some time later, at the next inspection, that the bad work was spotted by another mechanic. In the other case, the work was done by a mechanic and signed off by an IA who never saw the airplane. (He was based several miles from where the work was done.)

Beware of a low-time engine that is several years old or had a major overhaul several years ago. It is harder on an engine to sit idle than to work. In other words, a 150-hour engine that has acquired that time over a 5-year period is likely not as good as a 150-hour engine that got that time in a 2-year period. When I say "not as good," I mean it won't go nearly as long before you have to major it.

Cosmetics are not as important as mechanical condition. Pretty airplanes don't fly any better than ugly ones. (Has anyone ever seen a really ugly airplane? They're all things of beauty, aren't they?) Even so, good paint and interior tells you something about the owner and how he treats his property. Has the owner lavished loving care on it or let it deteriorate? Is it suffering from neglect? How long has he or she owned it? This might tell you something.

With respect to financing, you might find it advantageous to obtain financing other than on an aircraft loan in which the airplane itself is the collateral. The interest rates on aircraft financing are usually among the highest around. Many people use home-equity loans to buy airplanes. This results in lower interest rates and much lower payments (due to longer terms and lower interest). Do not put all your funds into the down payment, although you should put as much down as you comfortably can on the purchase price of the airplane. You should set some aside for contingencies.

Finally, if you plan to lease it to an FBO, flight school, or flying club, do not expect to generate a positive cash flow. The best you can hope for is to own, in a few year's time, a worn-out airplane that didn't cost you very much, and in the meantime your own flying got to be somewhere between cheap and free. Remember, renters don't treat equipment the way owners do. If you're lucky, the airplane will generate enough income to come close to making its payments, and when you add the tax advantages, at the end of each year you will have made a small paper profit. But you will have experienced a negative cash flow. And this is if you got a good deal with a reputable operator!

You should be aware of the fact that aviation (like many other fields of endeavor) abounds with horse traders and unethical business operators. In fact, aviation seems to attract more than its share of such unscrupulous characters, or "operators." People new to aviation should be aware of these con artists. There are self-proclaimed "experts" in this field who have little or no experience with the products or services they sell. If you do buy an airplane, I strongly suggest that you do not rush right out and overequip it, as so many first-time buyers are prone to do.

Probably the best way to go is shared ownership. With one or more partners to share the fixed costs of airplane ownership, your flying can be reduced to the operating costs only, plus the fact that your initial investment is so much less. Of course, you will have to bear your share of the fixed costs as well, but this will be substantially less than if you carried the load alone. The key is to have the right partner or partners. Do not become involved in shared ownership with guys who are overextended and unwilling or unable to come up with the cash when things need to be done. I once belonged to a club that had 17 members and one airplane. It may seem hard to believe, but availability was no problem! There were only 3 of us who flew the airplane. Thus, our flying was being subsidized by 14 other guys. It was much better than owning my own. In the more than two years that I belonged to that group, there were only two isolated occasions when I wanted to use the airplane and it was already scheduled by another member. And both of those cases were spur-of-the-moment deals. I put over 75 hours a year on that old Cessna 170 at $3.00 per hour (dry) and $10.00 per month dues.

I hope these tips have proven helpful. Remember, don't rush into the purchase of an airplane; they're easy to buy and hard to sell. On the other hand, generally, airplanes don't depreciate; they appreciate! (This can be a problem when you sell your baby and have to recapture all that depreciation you took off each year when you filed your tax return with Uncle Sam if you were justifying a commercial or business use.) Whatever you finally decide to do, good luck.

## II Should you take Fido?

Nearly half the households in the United States include one or more animals, and many of these households also include a general-aviation pilot. Arrangements must be made for the care of these pets before the human members of the household can take a trip of more than one day's duration. Do statements such as these sound familiar?

"I can't go. You'll just have to go by yourself. There's nobody to take care of Princess."

"What do you mean, the boarding kennel is 15 dollars a day?"

"I'd rather stay home. I can't stand the thought of being separated from my darling Tootsie."

Every pet-owning pilot should know that it is not only possible to take his or her dog or cat along in the aircraft, but that if done properly it is not particularly difficult to do so. However, proper precautions must be taken to avoid the sort of disaster described in Chapter 1.

I posed the question to Myrna L. Papurt, D.V.M., who is a graduate of the Ohio State University School of Veterinary Medicine and a long-time practitioner of animal medicine. She told me, "Sure, you can take 'em. You can certainly travel in light aircraft with nearly any dog or cat, and many exotic pets as well. The only thing is, you have to be sure to do it right."

When I followed this by asking, "Just what do you mean, 'do it right'?" she gave me a whole lot of pertinent information, all of which makes sense. Doctor Papurt has, along with practicing veterinary medicine, owned, trained and shown a large number of dogs, cats, horses, and exotics, and she has traveled extensively with many of her animals. I will here attempt to pass on some of her recommendations to those of you who are fellow pet owners.

First, and of paramount importance, Doctor Papurt told me, is to be sure that the animal will be welcome at your destination and that proper facilities will be available to care for it during your stay. Do not take Fido along on a whim, and then when it's too late discover that the hotel doesn't allow dogs. Some hotels and motels have just such rules while others maintain that they'd rather have your dog than you. After all, dogs don't smoke in bed and start fires, nor do they spill booze, coffee, or other beverages and stain the carpet, nor do they steal towels or other hotel property that isn't tied down.

Don't consider taking Fluffy before you inquire if your host is deathly allergic to cats. Better not take Rowdy if you're staying in a cramped apartment with small children who are afraid of dogs. On the other hand, if you have certain knowledge that your pet will be welcome and comfortable at your destination, by all means don't hesitate to take it along.

Pets in general-aviation aircraft are an entirely different matter from pets flying on an air carrier. Airlines require that all animals be confined to an animal carrier of specified dimensions and that all but the very smallest of these animal carriers be consigned to the baggage area. Most airlines have such designated equipment available for rent or sale. In a general-aviation aircraft one can, of course, do whatever one chooses. It is absolutely imperative that you choose safety for your animal. And step one is the use of a pet carrier, the system which has been working so well for so many years for the airlines.

A pet carrier is a wonderful idea. Doctor Papurt recommends such a portable cage for all cats and most dogs. A pet in a carrier cannot become frightened and bolt away, requiring an extensive search, or perhaps the loss of the animal altogether. It cannot hide under a seat where it would be difficult to retrieve. It cannot jump on the pilot, possibly with extremely unfortunate results to all concerned. It cannot urinate, defecate, or vomit on the aircraft upholstery. A carrier for one's pet shouldn't be too large. It takes the place of a seat belt for the animal, and if it is larger than required for the specific pet, the animal could well be injured bouncing around inside the carrier if turbulence should be encountered, or in the event of a rough landing. The pet should have enough room to stand up and turn around in the carrier, but not much more, according to Doctor Papurt. She asserts that a good rule of thumb is that the carrier need not be more than one and one-half times the length of the animal.

Doctor Papurt goes on to recommend that every pet in a carrier should be transported in the cabin with the passengers rather than in an isolated baggage compartment, such as the nose compartment of a twin-engine airplane. The environment in such an area is neither temperature or noise controlled, which may be okay for luggage, but not for a dog or cat. As is the case with everything in an aircraft, the carrier must be secured. A good way to do this if there is a seat available and the carrier is small (cat carrier or small dog), is to place the carrier on the seat and run the seat belt through the handle. For larger carriers, a seat may be removed, or the carrier may be secured in the baggage area behind the seats.

The plastic or fiberglass airline carriers with a wire door and ventilation holes or grates are superior to the wire cage types. Doctor Papurt mentioned several sound reasons for this. The solid (plastic or fiberglass) carrier prevents the pet from seeing strange objects and thus enhances the feeling of security. It lessens the likelihood of the animal being overtaken with fear, and protects it to some degree from extremes of temperature. And should the pet have an unfortunate "accident" the result is confined to a leakproof enclosure and not distributed throughout the interior of the aircraft, resulting in an extremely unpleasant environment for everyone aboard, an event to be scrupulously avoided at all costs. The desirable sort of carrier can be obtained at many pet supply houses, or through some airline offices.

The floor of the carrier should be covered with some absorbent material. For older pets, a towel or a blanket is very suitable, but for puppies, which are likely to chew and perhaps swallow pieces of cloth, newspapers are perfectly adequate. If the animal must be left unattended, the carrier should be equipped with a lock. Food and water dishes should not be placed inside the carrier, but rather offered to the pet at prescribed intervals.

Some dogs, especially large breeds that are well trained and accustomed to automobile travel, will readily adapt to riding in a general-aviation aircraft without a carrier. It is important to make certain that any animal that is not to be confined in a carrier have a temperament that is not easily incited to fear or panic, or the results of attempted air travel could be extremely traumatic indeed to both the pet and the people aboard, and to the aircraft itself, as I found out one time when I executed a missed instrument approach at night. The sudden application of climb power and pitch up of the Aztec I was flying caused our Doberman to defecate on the rear seat where she was restrained with a seat belt through a harness.

Doctor Papurt insists that no animal, even those not confined inside a carrier, should ever be at large inside an airplane. Every dog that is not in a carrier absolutely must be fitted with a comfortable, escape-proof harness and a collar with an identification tag and a leash. The leash should be attached to the collar at all times.

The harness is to be used with the seat belt. Every large pet supply house stocks dog harnesses in a variety of sizes. Papurt says the best ones for large and even medium dogs are the types of harness designed for hunting and tracking dogs. These are much more difficult for the animal to pull out of than the ordinary style. Very small dogs can be fitted with a cat harness, which is more escape-proof than the more common small dog harness. Several years ago, when I owned and regularly flew a four-place taildragger, my wife and I had a registered racing Greyhound with an obedience degree. We travelled with it so often that on the occasions when we didn't take it along, the tower controllers at our home base would ask where it was. We used the harness technique, and I can attest that it worked quite well for us. Without the benefit of Doctor Papurt's good advice, we blundered into doing just as she recommends here, probably because my wife is a very smart lady.

As for feeding the pet prior to a trip, this depends upon the pet and the length of the trip. Remember, pilots can't pull over to the nearest cloud and park while they take care of their pet's needs. Generally, it is best to withhold food for 6 to 12 hours before departing. This means skipping one meal for an adult animal and perhaps more for a puppy or kitten. This apparent cruelty can be rectified at the destination, and in the meantime the pet will not be particularly uncomfortable or smelly. Some dogs, especially older ones, should never have water withheld. It is a wise idea to take along the food to which the animal is accustomed to avoid upsetting its digestive system by a sudden change in diet. On the other hand, any potable water is usually satisfactory for a pet.

Just like us, pets should empty their bladders before traveling by air. With an increase in altitude, the bladder and its contents expand and passengers feel like they have to go even if they don't. It should also be noted that the animal's ears work just like ours, so it is a good idea to attempt to equalize the pressure by offering Fido a doggy biscuit every 500 feet or so of altitude change, and to make gradual ascents and descents.

An extremely important point stressed by Doctor Papurt is the fact that heat kills. She mentioned that owners are usually careful to see that Fido doesn't get cold, but dogs withstand cold temperatures much better than they do excessively hot ones. This is particularly true of the brachycephalic and heavy-coated breeds. Additionally, the smaller the dog, the greater the skin area it has in proportion to its size. The owner must beware of traveling in very hot weather with a hairy, short-nosed Pekingese in a carrier, else the owner may arrive at his destination with a dog dead of heat stroke. Bulldogs, Pugs, and other short-nosed dogs, even though their coats are not especially heavy, are also at risk, as are Peke-faced Persian cats. As a general rule, all dogs and cats should be kept out of the sun in a well-ventilated area.

Over the years Doctor Papurt has frequently been asked to prescribe tranquilizers for pets that are to travel by car or air. Usually she recommends against this procedure because a pet tranquilized enough to alter its behavior often loses bladder and bowel control. Such a situation may be very unpleasant for its fellow passengers and the individual who is required to deal with the pet at the destination. Mild sedation, however, can be a good idea, she says. She adds that any pet that experiences significant motion sickness in an automobile is likely to be at least twice as adversely affected in an aircraft. What else is new?

Cats almost never exhibit motion sickness, but dogs often do. Certain human drugs, given in appropriate doses, are applicable to administer to a dog and are readily available without prescription. The most commonly used of these drugs is, of course, Dramamine, which comes in a 50-milligram tablet. I was told that the easiest way to get a dog to swallow a tablet is to enclose it in a piece of meat, such as hamburger, or a slice of hot dog, and if the animal is first given a piece without the tablet, it will usually grab the next piece, which may well contain the tablet within.

The usual dose of Dramamine for a dog is 1 milligram per pound of body weight, so a dog weighing 25 pounds would get one-half of a standard tablet, repeated every four to six hours. Drowsiness is frequently a side effect of this drug, and, of course, simple drowsiness is desirable. As much as 2 milligrams per pound of body weight is perfectly safe for a healthy dog, and this much is likely to make it quite sedate. If a dog has any chronic medical condition or is taking any other medication, a veterinarian should be consulted before this or any other drug is administered. Good car riders will readily adjust to riding in an airplane.

All dogs that accompany their owners on trips should be current on all vaccines, and owners should be sure to take the vaccination certificates along with them, just in case proof is required by some authority someplace. There are, of course, special requirements for dogs that are traveling outside the United States. Dogs and cats being taken into Canada are required to have a certificate signed by a licensed, accredited veterinarian. This certificate must clearly describe the specific animal and state that it has had a rabies vaccine within the past 36 months.

The regulations regarding taking a pet into Mexico are somewhat similar to the Canadian rules, but when Doctor Papurt made inquiry, she was informed that the regulations may be subject to change without notice. To ensure that the traveler has current information and valid certificates, she suggested contacting the Mexican authorities a week or so before scheduled departure.

The question of whether or not to take your pet along when you fly rests on its welcome at the other end, and just how important it is to have it with you. Anyone who has a pet knows that there is a certain minimum amount of care required, and travelling by air with one's pet in general-aviation aircraft is certainly not particularly difficult nor is it more troublesome than caring for the pet at home, but the foregoing advice from Doctor Papurt certainly should be heeded.

Personally, my present dog is a Border Collie, and my friend Craig, a veterinarian, tells me that that breed is so smart, a handful of dog biscuits and a couple of hours of training and the dog could fly the airplane. He also says that with the sophistication of the automated glass cockpits the air carriers have today, they only require a crew of two: a pilot and a dog. The pilot's job is to feed the dog, and the dog's job is to bite the pilot if he attempts to touch anything.

# III  What does that do?

## Fear

"What are you doing that for?" "What does this thing do?" "Why are we tipping like this?" These are questions almost every pilot hears at one time or another from first-time passengers, or even from some regular but infrequent passengers.

It doesn't necessarily mean that you've got a nervous passenger on your hands, perhaps just a curious one. In either case, it is best to answer these questions fully, but in the most nontechnical language possible. Maybe the question is motivated by a desire to replace superstition and fear with science and knowledge. At least it is in this context that the pilot should answer whatever questions the passenger asks. It has been said that fear kills reason, and I'm sure this is an accurate statement. It is the unknown that people fear, and flying machines are filled with gadgets that are totally unknown to the average individual. To the uninitiated it seems that the machine itself is doing something impossible—it is defying gravity, and everybody knows you can't do that. There are some people who fly regularly as passengers and every time they arrive safely at their destination, they have the distinct feeling that they've cheated death one more time. I have actually seen people get out of an airplane and kiss the ground because they are so happy to once again be down on dear old Mother Earth.

The people who operate the air carriers know that what causes nervousness among passengers is the helpless feeling of sitting in a long tube with absolutely no control over their own destiny. Some yo-yo up front is driving the thing, and they have no idea what the pilot's doing to keep it up in the sky. These people know that the slightest mistake on the part of the pilot will result in that huge machine falling out of the sky and crashing to earth, killing all aboard. They feel that they are taking their lives and putting them in the hands of they know not whom to do with they know not what. They expect the machine to crash and burn. They expect to die.

We in general aviation can at least explain lift, weight, thrust, and drag to our first-time passengers so they'll have an inkling as to why the thing doesn't fall out of the sky. And for the benefit of the passenger who is lucky enough to sit up front, we can explain what each of those mysterious gages on the panel does.

I recall a man, a U.S. congressman, that I used to fly around on business trips. This was a man who had spent a considerable time on air-carrier airplanes apparently without a qualm, perhaps because he knew there were a minimum of two qualified pilots up front, but he seemed to be quite concerned in a light twin with just me at the controls. Do I inspire confidence or what? This gentleman expressed his concern by repeatedly asking what action he should take if I suddenly became incapacitated while we were alone in the airplane. I must have looked as though I was about to expire at any moment. I attempted to put him at ease by explaining how the controls worked, and how to work the autopilot. He then asked me to explain what each of the instruments does, and how to navigate to an airport. Fortunately, as an instructor I was able to replace his superstition and fear with science and knowledge. He gained confidence and finally came to enjoy flying in a light twin. In fact, he went so far as to tell me that if his busy schedule permitted, one day he would undertake flight instruction. This case was no doubt a combination of the two, both fear and curiosity.

Fear of flying has been proven to be not related to fear of heights. It is strictly fear of the unknown and should be easily treated with education—should be, but isn't. This fear is so deep-rooted that in many cases simply explaining the scientific principles involved is not enough. When the emotions take over, logic goes out the window, and fear is a very powerful emotion. In these extreme cases, the best solution is to substitute another emotion, if possible one of equal or greater power. Try pleasure. If flying can be made a pleasurable experience, perhaps fear will dissolve. How to make it seem fun to the reluctant passenger? Put him or her to work. If they feel needed, people's minds are too occupied with whatever task is thrust upon them that they forget to be afraid. Don't forget, it is a lack of control which causes people to be anxious about their welfare, and if they have something to do, it at least gives them a modicum of control, or at least lets them think so. Even if it is only tuning a communication radio, it is something. When people become involved and are interested, they lose their fear. It is much the same as handling student pilots who are prone to motion sickness. Generally, when the instructor turns the controls over to these students, they become so involved in manipulating the airplane that they forget to become airsick. So it is with frightened passengers. They become so occupied with their assignments that they forget to be afraid. They are is simply too busy.

Somehow, being occupied seems to remove the insecurity of the unknown, and it is this lack of security, this uncertainty that manifests itself in the fear of flying. If you can get these people involved with some kind of hands-on task, they will feel that they have at least some control over what's happening, and their fear will likely dissolve.

## Curiosity

On the other hand, those questions such as, "What does that do?" or "How does that work?" are likely to be prompted by simple curiosity. This provides pilots with occasions to show off their knowledge. Everybody likes to feel that they are special, and pilots are no exception. To a greater or lesser degree, we are all flattered by being asked to explain something we know and understand which the other person doesn't. This situation presents the pilot with an opportunity to enlist another aviation enthusiast and should most definitely not be wasted. After all, how often does one get a chance to show off knowledge and skill while at the same time sharing pleasure in aviation?

By carefully avoiding insulting the listener's intelligence by oversimplification, while refraining from using technical jargon, pilots can pique the interest of their passengers. Pilots can point out interesting landmarks outside the airplane and explain how the navigation receivers work. They can show how banking the airplane makes it turn and how the elevator changes pitch. Pilots can explain pitch, roll, and yaw while demonstrating all three, even if they aren't instructors. They can even encourage their passengers to try maneuvering the airplane themselves. Many people will tend to be somewhat timid about trying this, and they certainly should not be forced or made to feel any pressure, but rather gently encouraged. The key here is to emphasize the fact that all control movements must be very smooth and easy. No jerky movements allowed! Lacking coordination, the passenger is likely to slip or yaw the airplane, so it may be necessary for the pilot to slightly assist with yoke or rudder pressure.

If pilots don't know or are unable to give a factual and accurate answer, they should say so rather than attempt to bluff. They can tell the curious passenger that the question is better answered back on the ground. Then, after landing, they can look up the answer together. Having questions answered this way, the passengers may never find out that the pilot was unsure of the correct answer. The passenger will feel a greater sense of accomplishment, having participated in the process of having looked up the answer rather than simply been told the answer, and the result is that both parties learned something.

It is easy to convince the curious passenger that the world's worst classroom is the cockpit of an airplane with the prop going around, that all flight instruction really takes place on the ground, and that the airplane is simply a learning tool used to prove what has already been learned on the ground. Passengers' curiosity regarding whatever they wanted to know is then completely satisfied. And, of course, if they're lucky, the pilot has helped to recruit a new student for the flight school.

## The unnatural

Flight is one of the most unnatural activities in which human beings can engage. The sky is not our natural environment. Even those who sail the oceans are not in quite as strange a situation as those who traverse the sky. We know that as long as humans have walked the earth, from that time beyond which the memory of humans runs not to the contrary, we have always envied the birds as they soared through the sky. And we have longed to join them, mostly without success—witness Icarus and his disastrous effort. In fact, all efforts failed until 1908 when the brothers from Dayton finally succeeded.

But what happens when we do conquer the sky? We find ourselves in a totally unnatural environment. Nothing about flying is natural. It all has to be learned. Everything about flying is a learned experience. And the better we learn it, the more successful we are. Through constant practice and repetition, we acquire a set of conditioned responses which enable us to successfully operate in this unnatural environment. And these responses are quite complex, which is what makes learning to fly such a difficult proposition. Way back when I was a student pilot I was told, "Learning to fly is the most difficult thing in the world, but doing it is the easiest." And I've never seen anything since to contradict this statement. The old CAA (Civil Aeronautics Administration—the predecessor of the Federal Aviation Agency) *Flight Instructor Manual* defined coordination as the use of more than one motor skill simultaneously to achieve a single, desired result. And in the final analysis, that's exactly what flying an aircraft is all about. This kind of response can only be realized through constant repetition and drill. That's why flight instructors have their students go through each procedure and maneuver over and over again until they get it down perfectly, and then go over it some more. It then becomes an ingrained habit, a conditioned response. And it is this series of conditioned responses which enable us to exist in that unnatural medium, the atmosphere around us.

# 8

# Tips on flying advanced equipment

## I Checkout

Theoretically, pilots who are qualified by virtue of their certificates and ratings may conduct a self-checkout of any aircraft for which they rated. But in most cases this is not practical. Practical considerations throughout the aviation industry are governed more and more by the insurance carriers as airplanes and equipment become more complex and thus demand a greater degree of expertise on the part of the pilots. Our equipment has become more expensive and our liability greater, so the insurance carriers have a greater stake in what we do and how we do it.

In my lifetime I have had many occasions to check myself out in new and different-make-and-model airplanes. The ideal, of course, is to have a qualified instructor who is familiar with the specific equipment under consideration conduct the checkout, but this is not always possible. In the very simple equipment we used to fly not too many years ago, all that was required was to take off, climb to a reasonable altitude, do quite a bit of slow flight at minimum controllable airspeed while feeling out the airplane, perform a stall series, a couple of steep turns in each direction, and come back for a few landings. Before taking off, I would sit in the airplane on the ground and get a firm grasp on the "sight picture" of what the landing attitude should look like, and when doing the stall series I would concentrate on "landing attitude stalls."

This system worked fine with the Mooney Model M-18, the Mooney Mite (a single-place airplane that had to be flown alone), the Luscome, the Cessna 170 (and even the 180 and 195, from which there is zero forward vision on the ground) and most of the airplanes of that era. After all, we didn't have any such thing as a POH (*Pilot's Operating Handbook*), and rarely even a check-off list. Then, several years ago the FAA came out with two excellent ACs (advisory circulars), AC 61.9 and AC 61.10, guides for instructors checking out pilots in new and different (to them) equipment. These two advisory circulars covered the recommended checkout procedures for complex singles and twins, respectively, and they can be used as guidelines for the pilot conducting a self-checkout and for the instructor pilot checking out another pilot unfamiliar with the type.

Unlike the system described above, today's sophisticated airplanes require a bit more than "kick the tires, start the fires, and off we go." I have enjoyed quite a bit of success checking myself, and other pilots, out using AC 61.9 and 61.10, although in each case I would have preferred to have a competent instructor to take me through the procedure. The first step, of course, is to read (study) the manual and learn the various systems in the airplane, and such important things as the crucial airspeeds and weight and balance data. When I check out another pilot, I give him or her a test on that stuff, some of which must be committed to memory and some of which need only be handy for quick reference.

A careful and thorough walkaround inspection, using the checklist comes next. Then, before attempting to fly the airplane, it is important to sit in the pilot's seat and go over all the equipment (avionics—knobs and dials), controls, etc. Since all modern airplanes have comprehensive checklists in the manual, follow this list carefully and completely through the startup procedure. (Don't use the abbreviated checklist found on the lower panel of most Piper aircraft, or in printed form.)

From this point on, if the procedures suggested in AC 61.9 (or .10 as the case may be) are followed, a successful checkout will follow. It is vitally important to learn and know the systems in each airplane you fly and the procedures by which it is flown.

Best of all are the courses offered by the manufacturers or those offered by a manufacturer authorized facility. I have been fortunate enough to have had the opportunity to go through type-specific schooling offered by all three, Beech (Model 60 Duke, and Model 90 King Air), Piper (Model 601P Aerostar), and Cessna (Model 421), plus the Aero Commander 690, and I can tell you they are all superb learning experiences.

# II Initial training in complex aircraft

Traditionally, pilot training in airplanes has started in simple fixed-gear, fixed-pitch airplanes. After the trainees have achieved the status of private pilot and acquired their certificate, they may go on to check out and qualify in more sophisticated, complex airplanes. This is how the vast majority of pilot training is accomplished. The student trains in a Cessna 150-152, and immediately after acquiring his or her certificate, checks out in a 172 Skyhawk, or a Cherokee. If trained in a high-wing airplane, during transition to a low-wing airplane, he or she must learn about fuel pumps and fuel management and perhaps manual flaps as opposed to electric flaps. Back when I was trained in the old Army Air Corps (before we had an Air Force), we started out in a PT (primary trainer—65 hours), moved on to a BT (basic trainer—another 65 hours), then to an AT (Advanced Trainer) before transitioning into an operational fighter or bomber. The military still does it this way, but their pilots start in heavier, faster, and more complex airplanes.

That's pretty much how it is usually done. However, on rare occasions, particularly when a student pilot buys an airplane, an instructor may be asked to undertake to train a beginning student in a complex airplane with a variable-pitch, constant-speed propeller and retractable landing gear, sometimes to even train a primary student in a twin engine airplane. When such a request is made, it poses a unique problem for the instructor, although not an insoluble one.

Over a career as a flight instructor spanning almost 50 years, I've had the experience of training students from the very beginning in Beechcraft Bonanzas, Mooneys, and even in a couple of twins. All this really does is require that students have more than the usual experience when they complete their training. This kind of training, although difficult, is by no means impossible. As is the case with virtually everything else, when students learn to do anything one step at a time, the end result is that they have it all. Broken down into a series of simple tasks, although not easy, it becomes a great deal less difficult. I had an algebra teacher way back in the eighth grade who started the course by putting a long, involved formula on the blackboard. It was totally incomprehensible to the entire class. The teacher told us that in 12 weeks, when we got to page whatever in the textbook, it would be simple. He left this string of numbers and symbols on the board, and sure enough, 12 weeks later, when the class had reached the specified page in the textbook, we all understood it. The same principle applies to training in a complex airplane. After each simple step has been mastered, it all comes together.

As each system in the airplane is tackled, the instructor can use the teaching techniques set forth in the *Aviation Instructor's Handbook* (AC 61.14). For example, in teaching the purpose and procedure of the adjustable prop, the technique of transfer of existing knowledge can be used. If the student drives an automobile, the similarity of function between the constant-speed propeller and the transmission in the car can be explained to the student. It may be somewhat of an oversimplification, but it is effective.

Similar to the fact that the low gear in a car enables the machine to develop the power to overcome the inertia of the standing auto and get it rolling, a low-pitch prop turning at high speed enables an airplane engine to put out its maximum power for takeoff and climb. When the automobile is rolling along, it is shifted into high gear for efficiency and speed, similar to what the propeller does with the airplane when it is set up for cruise. It is "shifted" into the high-pitch, low-speed mode for efficiency and speed. The prop is turning slower and grabbing a bigger bite of air with each revolution. The British term for propeller is "airscrew," and this is definitely a more realistically descriptive word for that thing out in front that pulls the airplane along through the air. All we're doing when we adjust the pitch is changing the size of the bite or slice of air that it is taking as it revolves.

Just as attempting to start a car rolling in high gear causes the engine to "lug" (and probably quit), so attempting to take off and climb with the prop in high pitch will overwork the airplane engine, likely causing severe damage. Using this technique of transfer of knowledge, the instructor can get across to the student the function of the adjustable propeller.

The landing gear, however, is another matter entirely. It is either up or down, and it is fairly simple to teach anyone to use a checklist. Of course, anyone can fall victim to distraction and make the classic mistake of a gear-up landing, but it is no more likely with the beginning student than it is with the experienced pilot, perhaps even less so, for with the experienced pilot complacency may set in. They say there are those who have landed gear up and those who are going to. Well, I haven't done it yet, but I'm not here to say that someday I won't. I have tried to do it twice, and both times I was saved from the embarrassment of that short step down by an alert controller. Both were classic cases of distractions demanding my attention and diverting my attention from the normal routine prelanding checks. We teach primary students to use flaps, don't we? Well, if they can be taught to extend and retract flaps at the appropriate times, so can they be taught to raise and lower the undercarriage at the appropriate times.

Way back when insurance companies were more generous with their pilot warranties, covering students flying solo in almost anything, I trained two student pilots (a pair of young brothers) in an Aero Commander 500 series airplane from the very beginning. Their father owned the airplane and he thought they ought to learn in the airplane they would be flying. When they finished they both got Private Pilot Certificates and were rated for Airplanes-Multiengine Land with no single-engine privileges whatever. I don't claim this was easy, but it wasn't as difficult as one might think. I don't know if either or both of them ever acquired single-engine privileges, but I do know that one of them went on to become an air-carrier pilot for a major airline. More recently I had a student who had just bought a Cessna 310 come to me and insist that he get his primary training in his own airplane. He, too, wound up with multiengine privileges only.

Today, the insurance carriers have much more stringent requirements than they did when I trained the two brothers in the Aero Commander twin. My associate, Charles Sova, recently had the experience of training a primary student from scratch in a Bonanza worth well over 150,000 dollars, and the additional premium over and above the regular insurance to cover him when he was solo in the airplane was 5,000 dollars. Even so, since he owned the airplane with a partner, he opted to pay the extra premium just so he could train in his own airplane rather than one of the school's primary trainers, which would have been a much less costly affair. Of course, once he finished his training and got his certificate, the insurance premium went down.

In spite of Chuck's efforts to talk the student into doing his training in one of our school airplanes, the guy insisted on doing it in the airplane he would be flying. Because the airplane goes so much faster than the conventional light trainer, it required about a dozen cross-country flights to meet the minimum cross-country time requirement for certification. At first, Chuck thought about having him do the training with the gear down all the time, but since this would not duplicate the student's "real-world" experience, that idea was discarded. Next, Chuck considered doing everything in slow flight, but, again, that idea was discarded for the same reason. Finally, deciding that since anybody can (and many do) land gear up, both experienced pilots and neophytes, Chuck merely pounded on the use of the "GUMP" check. Since he was responsible for his student, Chuck says that when the student was out solo, he was prepared to throw himself in front of the crash truck as it was on the way to the site of the inevitable wheels-up landing.

Chuck Sova is the very best flight instructor I have ever known. (In well over 50 years of critically evaluating flight instructors, I've never seen a better one.) Although he has trained a couple of pilots in complex aircraft, he is of the opinion that they would be much better advised to train in a light fixed-gear, fixed-pitch trainer, and then transition to the complex airplane on the theory that it is substantially easier to do it that way than to start from scratch in the complex machine. I'm not sure I agree with that procedure. After all, we have to give the customer what he insists he wants, and if one follows the "one step at a time" approach, it works out okay. Of course, it does take longer, but when you add the transition time to the original training time, it isn't all that much longer or costly (except for the insurance factor and the fact that the operating cost of the complex jobbie are higher).

As a basic rule, the larger and heavier the airplane, the easier it is to fly and to land. It is also likely to be more stable than the small single-engine trainer. Thus, in some respects it is easier for the primary student to learn to manipulate a complex airplane. Although it is likely to be going quite a bit faster than the little two-place trainer, it is also likely to be less quick to respond to control input. It's all a matter of training the student to stay ahead of the airplane (see Chapter 5).

Of course, the student trained in the Bonanza or Cessna 310 is going to go a lot farther on his solo cross-country expeditions in order to get in his required 10 hours of experience than is the one putting along in his 150, 152, Tomahawk, or Cherokee. This means that the instructor must be particularly careful in seeing to it that the student's navigation skills are adequate for the job. Everything is happening faster and this places an additional burden on the student to stay ahead of the airplane. And at the faster speed it won't take nearly as long to get seriously off course on a VFR cross-country trip. On the other hand, such an airplane is likely to have some of the more sophisticated navigation equipment, making it even easier to stay on course.

Learning in this kind of airplane requires that the students thoroughly know and understand more systems than the usual trainer has. They must understand just how that propeller works, and if it is a twin with full feathering props, they must learn and understand the principle of feathering and unfeathering and synchronizing the props. They must know the landing gear system and the emergency extension system. And they must acquire all this extra knowledge at the same time they are learning the basics of manipulating the airplane, just as the students in the little trainers must.

Some of the major air carriers have begun to institute a program of ab initio cockpit crew training in which they take the students who have tested well and show a lot of aptitude but have little or no experience and train them to fly (in the simulator) some of the most sophisticated equipment in the world. The jury is still out on this procedure, but I believe it has certainly been worth a try. It may very well be the wave of the future insofar as training air-carrier pilots is concerned (see Chapter 15).

Military training is accomplished in very sophisticated equipment. Of course, military flight students have the advantage of working at it full time. They live, breathe, and eat flying as they go through one of the most structured curricula in the world, spending a solid year working half a day in ground school and half a day on the flight line.

Experience has demonstrated that just as soon as student pilots receive their Private Pilot Certificate, they will usually rush right out and check out the next airplane up the line—from a 150 or 152 to a 172, thence to a 182, and so on. If they have trained in a Tomahawk, they immediately move up to a Warrior, then to an Archer, then an Arrow, etc. This is the track followed by most renter pilots. However, the case of the owner-pilots is somewhat different. If they train in a conventional trainer and then buy an airplane, that may be the only thing they ever use after becoming certificated. On the other hand, if they bought their airplanes before they start training, and some do, then it just makes sense to train in what they'll be flying. On extremely rare occasions, students may learn in a borrowed friend's airplane, in which case they've trained in the one that's available.

Whatever the reason, it does sometimes happen, and it's not the end of the world. It can be done! I've even heard of people who got all their training in a floatplane, and the only privileges they have are in a seaplane flying off the water.

One of the very best students I ever had did his primary training in a CE 172. He did all his advanced training in Pipers (Cherokee 140, Arrow, and Aztec). After he acquired his CFI certificate, he went to work at our flight school. One day I asked him to ferry a CE 150 from one of our facilities to another, and he replied, "I've got over 300 hours, but I've never been in a 150!" I told him to just go ahead and do it! I also had a student owner one time who trained in her 172, and when it was time for her private checkride, her airplane was in the shop, so I sent her to the examiner in a 150. She loved it!

# III  Know your systems

A substantial portion of the training effort with respect to large aircraft is devoted to the various systems in the specific airplane. As we in general aviation move up into more and more complex equipment, it becomes ever more important to thoroughly know all the systems in the airplanes we fly. Not only must we know what each item does, but why it does it, just how it works, and what makes it do whatever it does. And this is an area that is occasionally brushed over, or ignored altogether by flight instructors.

We had an FAA inspector, now retired, in our district who devoted almost his entire oral portion of the multiengine flight test to having the applicant describe in excruciating detail just how the feathering and unfeathering of propellers works. Many pilots and instructors don't think it is important to know this. After all, all pilots need to know is that when they want (or need) to feather the prop on a dead engine, it is necessary to yank the prop knob briskly into the feather position before it has an opportunity to wind down enough for the lock pins to activate and prevent it from going into feather. I do, however, think it is important for pilots to know just why it is necessary to briskly pull the prop control to the feather position. If they know why they're required to do it, the pilots are more likely to be sure to do it right.

I recently received a letter containing an excellent example of the advantage of knowing the systems in your airplane and the proper use of your equipment. A good friend of mine, Steve Hess, who flies a Model 33 Bonanza out of Rochester, New York, had quite an adventure not long ago. Mr. Hess has built a moderately successful business for himself and is in a position where he can afford to own and operate a really nice airplane. He can also afford to equip it very well with a great many useful options. You might say his airplane has all the whistles and bells.

At any rate, the adventure with which we are here concerned occurred while he was en route from Rochester, New York, to DPA (West Chicago—DuPage). The purpose of the trip was to drop his airplane off so that he could have his KLN90A upgraded to a B model with an Argus moving map. Because of panel limitations, this would also necessitate the exchange of his Stormscope WX900 for a Strikefinder. While it was there the airplane was also scheduled to receive an annual inspection.

The weather on the day of this trip (it happened to be in mid-February) was somewhat grim with blowing snow, low visibility and ceilings, reported turbulence, and icing conditions, typical winter weather in the northeastern United States. It was IMC almost the entire length of the trip.

Steve knew that he had carefully checked the fuel caps at the time of his preflight inspection, and they were locked and secure. However, as he kept looking at his wings for the first indication of the accretion of ice, he noticed a wisping of fuel (visible vapor) around the right fuel cap, and he knew he was losing fuel. When he first noticed this he was about a half hour into the flight and was running on the left main tank. Steve decided to switch tanks on the theory that it would be better to burn fuel than to permit it to be sucked out in the form of visible vapor. He also believed that when he had burned off a few gallons and the remainder in the tank got lower, the fuel leak would stop. He had started out with a total of 104 gallons, including what was in the tip tanks. If there was no fuel at all in the right main tank, he would still have 67 gallons (37 in the left main and 15 in each tip), giving him over 4 ½ hours of endurance for the 3-hour and 20-minute flight. Additionally, all along the route of flight there were airports every few miles in the event things should worsen and a landing become necessary.

Therefore, he switched tanks. Mr. Hess is nothing if not cautious, so he made a note of the reading from the fuel totalizer, in his words, "another great peace of mind piece of gear." He says he writes things like that down on his lapboard because when the pressure goes up, the mind can slip on a person. He had burned 8 gallons from the left tank since takeoff, and the tips were full with 15 gallons each.

His plan was to burn 25 gallons from the leaking right tank. Steve mistakenly believed that the slight wisp he observed flowing from the cap would amount to perhaps as much as 5 gallons, maximum. Checking the right fuel gage and noting that it indicated $\frac{7}{8}$ full, he leaned back and relaxed. An hour later the fuel gage was still reading near the top, and as he put it, "I realized the leaking 'O' ring was causing the tank to be pressurized, lifting the rubber bladder and rendering the fuel gage useless."

He kept his eye on the very accurate Shadin Fuel Totalizer, another very useful gadget, and noted that he had used only 20 gallons from the leaking right tank. His plan, based on his estimate of the amount lost, was to continue on that one for another 15 minutes and then switch. Suddenly the engine quit and an awful silence descended on the airplane at 4,000 feet in IMC. He had stayed low because of ice and headwinds. As we all do when the adrenaline starts flowing in a quiet airplane, Steve began to move very fast. With tanks switched, throttle and mixture set, finger on the fuel pump switch, it took just a few seconds for that wonderful noise to return. He had lost a mere 100 feet of altitude, and almost lost his half-digested breakfast. He reset everything, started fuel transfer from the tip tanks, and completed the flight to his destination.

Steve points out that a slight cap leak can be deceiving. He had lost 17 gallons. However, the real problem would have been if it had happened at night or with a high-wing airplane, using the same kind of fuel bladders which enabled the gage to continue to read "full." Of course, the fact that Steve had more than enough fuel certainly didn't hurt. Despite all this, the fact remains that Steve Hess' knowledge of his aircraft systems was the best thing he had going for him. By the way, a majority of pilots are very reluctant to run a tank dry, on the theory that the last dregs at the bottom of the tank might be contaminated. I believe this is an erroneous assumption. All the fuel is always drawn from the bottom of the tank, isn't it?

Even a relatively conscientious pilot can sometimes fail to adequately learn the systems in his airplane. I know of a case in which the owner of a Cessna 310R who, after having owned and flown his airplane for quite some time, came to an instructor who specializes in recurrent and refresher training in complex singles and twins. After an hour or so of discussion on the ground, they went out to fly the airplane. The instructor's policy on their first flight together is to observe and evaluate the pilot's habits and skills before attempting to administer any real training. He told the aircraft owner to just go ahead and do what he normally does. The owner-pilot just climbed into the airplane and started to strap it on. The instructor asked him if he had preflighted it before he taxied over from the hangar to the restaurant where they had met. The reply was in the negative, so the instructor had the pilot get out with checklist in hand and start his walkaround inspection. He drained and checked the fuel from four sumps and when asked how many drains the airplane has, he replied that there are four. Knowing better, the instructor showed him two more that he didn't even know existed. The pilot showed the instructor his checklist, which had been made up by the former owner of the airplane, and it listed only the four drains. The manual, of course, lists all six, but the owner-pilot had never bothered to thoroughly learn all the systems in his airplane by studying the manual and its supplements, yet he had a reputation as a careful pilot. He had simply been using the abbreviated checklist that he inherited with the airplane.

There are many other examples demonstrating just why it is so important to thoroughly know and understand the systems in the equipment we fly, but these should suffice to make the point. In case they haven't, be advised: know the systems in the airplane you fly. That knowledge could save you from serious embarrassment. It could even save your life!

Of course, it is not only important to know the systems in the airplane you fly, but the equipment itself. Some modern equipment is so complex and manuals so poorly written that a new owner just has to play with it for quite a while until he or she learns how to use all it has to offer.

# IV Use everything available

It really bothers me to fly with pilots who permit much of the expensive equipment in their airplanes to just sit there on the panel doing nothing at all. This is an entirely different situation from that which exists when pilots buy airplanes and never learn how to use all the goodies that were installed in the airplanes when purchased. That one is simply a matter of knowing the equipment and the systems of the airplane. Here I'm referring to the pilots who know how to use all that stuff but who permit some of it to be lazy and just be there, perhaps not even turned on.

Personally, when I fly, I want every single needle in that airplane pointing to something useful. I have often sat in an airplane and watched a pilot fly down an airway without having any idea just where he was along the way. He permitted his number-two VOR to just lay there sound asleep rather than tracking his progress en route. My policy is to rotate the OBS 10 degrees ahead (full scale deflection) of a station off to the side, and when it has marched across the face of the instrument until it reaches full scale past the station, repeat the process. This way I always know pretty well just where I am. Of course, with a DME, LORAN C, or GPS, I can also track my progress. When using this method of ascertaining my position on the airway, I have my number one on the station ahead and the number two on the last one behind, regardless of the changeover point. This way I know beyond a doubt that I'm tracking straight down the center of the airway.

And on the subject of DMEs and VORTACs, an appalling number of instrument-rated pilots were never taught by their instrument instructors how to shoot a DME arc. I realize that with the almost universal Radar coverage we have today and all the other goodies that help us get lined up for the approach, the DME arc is not nearly as useful a tool as it was just a few years ago. However, there are still lots of places that can only be reached by means of the arc.

Another piece of equipment the use of which is occasionally not properly taught is the autopilot. When pilots buy airplanes equipped with autopilots and retain the services of an instructor to check them out in their new airplanes, the instructors usually have the pilots fly manually throughout the checkout process. This may result in the owner-pilot never getting thoroughly checked out on all the features of autopilot. This most frequently happens when someone buys an airplane that has an autopilot already installed. The new owner then decides it is time to undertake instrument instruction, and throughout instrument training he/she is not permitted to use the autopilot. Of course, he or she is expected to read the manual, but may not do so, or may not have received one with the airplane's paperwork. On a flight test the applicant may be asked to demonstrate the use of anything in the airplane. If it is installed in the airplane, it is fair game. And on more than one occasion when I have asked owner-pilots to show me how the autopilot works, I have been told, "I don't know how. Nobody showed me that." Of course, the pilots could have studied the manual and taught themselves, but some people don't seem to realize that such a thing is possible. It is always best to have a qualified instructor teach the use of strange or new and different equipment, but if a good manual is available, properly certificated pilots should be able to check themselves out, and this applies to airplanes and avionics and other equipment.

Although everybody should be capable of flying their airplanes themselves, there's no use in having an expensive piece of equipment such as an autopilot in your airplane if you don't use it, at least some of the time. It's a great work saver, and it really takes the load off the pilot when things in the cockpit get really busy. I'm certainly not advocating that pilots should abandon their responsibilities and always use the autopilot as soon as they're off the ground, but as long as your skills at manually flying the airplane are up to speed, you might as well use it. Then there are those, quite a few, who manually fly their airplanes so rarely and are so totally dependent on their autopilots that if the thing went to sleep, they'd be totally lost. This situation is, of course, much worse than if they never use it at all.

When I say I want every needle in the airplane pointing to something useful, that includes the ADF, which most pilots today use primarily for listening to the ball game. I try to keep my ADF tuned to some station near my route of flight as a further check of my progress. And there are still a great many places where the NDB approach is the only means of getting in under IFR. There are still even a few route segments that are defined by ADF bearings.

Did you know that an ADF can be used as a sort of poor person's Stormscope or Strikefinder? If there's really heavy weather with thunderstorms boomin' around out there, if you tune your ADF receiver to the lowest frequency on the dial, whenever there's a bolt of lightning anywhere nearby (within a few miles), the ADF needle will swing around and point toward it, thus offering a clue that may help you to avoid the very worst of the bad weather. Of course, a really active cell with lots of lightning will really drive that needle crazy!

And on the subject of ADF, those of you who are instrument rated, were you taught to leave the audio on (turned down low) throughout an NDB approach? By the way, although we use an ADF to execute an NDB approach, it is called an NDB approach because approaches are named after the ground equipment used. Anyway, since the ADF doesn't have an "off flag," the only way to know it is off the air is if there is no audio identification signal. Suppose the station goes off the air right in the middle of an approach? Without the audio, the pilot would have no way of knowing he wasn't receiving it. (This is one of those "what ifs?" covered in Chapter 16.) The needle would continue to point straight ahead, and whatever wind drift there was would take the airplane off the final approach course into who knows what, and another CFIT (controlled flight into terrain) might result. With the audio quietly chirping away with the station identifier, you know the station is transmitting and your receiver is working, so if you're skillful enough to do so, you can complete the approach to a successful conclusion.

Another technique rarely taught today is the DME arc method of alignment with the final approach course. I know what you're thinking. With the almost universal radar coverage we now have, who needs it? Well, let me tell you, there are still lots of places without radar coverage for approaches, and several of them have airports which use the arc. As a pilot examiner, once in a while I would ask an applicant whose airplane was equipped with DME to explain the technique of using the DME arc. (Whatever equipment is in the airplane is fair game for an examiner. The examiner can require the applicant to demonstrate or at least explain its use.) The majority of applicants I saw had never been taught how to shoot the arc, possibly because their instructors hadn't been shown by their instructors. If you are instrument rated and fly an airplane equipped with DME, for heaven's sake, find yourself instructors who know their business and get them to show you how to shoot the arc!

There's something all pilots, whether instrument rated or not, can do to make flying safer and more efficient, and that is they can take advantage of VFR flight following (see Chapter 11). I want all the eyeballs I can get watching me, so even when I'm just flying around the local area, I ask the nearest radar controller for advisories. Even if you are a VFR-only pilot, if you request flight following on every cross-country trip you take, it will do at least two things for you: one, it will help you avoid traffic, and two, it will give you an excellent opportunity to become comfortable in the ATC system. Always be polite and reasonable with your requests. After establishing contact with the facility, start off with, "Traffic and your workload permitting, I wonder if you would be kind enough to issue me advisories? I'm (give your location and altitude)." Then, before leaving that sector, ask, "Sir (or ma'am) your workload permitting, request a handoff to the next controller down the line." This way you'll be in contact with ATC all the way to your destination unless IFR traffic in one sector or another is too heavy for the controller to add you to his or her workload.

Be advised, however, that the controller's only real duty is to keep known IFR traffic separated. That's his or her first and only real priority. The burden is still on you to look around. I've seen lots of traffic that the controller failed to call out for me. Also, conversely, I have failed to spot lots of traffic that was pointed out by ATC! It just seems to work out that way.

# 9

# Ice, thunderstorms, fog, and other weather stuff

As a pilot examiner, I was frequently advised by instrument applicants taking their checkrides that they intended to hold themselves to higher-than-published minimums until they gained X amount of experience. You know what I mean, "I'll hold myself to 600 and 2 on an ILS and 1000 and 3 on nonprecision approaches for 6 months. Then gradually work my way down to the published minimums."

My answer was invariably, "Please don't tell me that. I don't want to hear any such thing. After all, if I issue you the rating, you are authorized to go right out and fly in the stuff on the gages right down to the published minimums and I must assume you will do so. Remember, every approach you executed during your training was to the published minimums and you'll never get any better unless you do it. As long as there are no blowing rocks and boulders or ice out there, you should be able to fly, in smooth air and without ice, in whatever weather you encounter. And you rapidly lose your instrument flying skills unless you keep them honed by constant practice."

Meteorology is the most inexact science of all. And weather forecasting is virtually never really accurate. How many times have you found exactly what you were told to expect? None, that's how many. I've been told to expect CAVU (ceiling and visibility unlimited), and then seen every kind of weather there is except good. For as long as I've been an aviator, people have been asking me, "What's the weather going to be tomorrow?"

My standard answer has always been, "Call me in the morning and I'll look out the window and tell you."

The two things instrument pilots (or any pilots for that matter) must most concern themselves with are ice and thunderstorms. Therefore, they are being treated together here.

# I Ice

For the first time in their aviation careers, the new instrument pilots become more aware than ever before of that little round dial that tells them what the ambient air temperature is—the outside air temperature gage. When there is moisture in the air, and you're flying at an altitude above the freezing level, you can pretty much count on picking up some structural ice. What most pilots don't realize is that the classifications of icing are based on the rate of accretion. In other words, trace doesn't mean you have just a trace of ice on your airplane, but just how long did it take to become a trace? The same criterion applies to light, moderate, heavy, severe, extreme, etc. If it takes 15 minutes to pick up enough ice to even be aware of it, and you're traveling at the rate of 3 miles per minute, the ice may not be a very great concern. On the other hand, if you pick up a quarter inch of structural ice within a couple of minutes, you are likely to already be in deep trouble.

Ice is weird stuff. We're told that there must be visible moisture present in the air for ice to become a problem for pilots. Don't believe it! It may be true of structural ice, but it doesn't necessarily apply to induction ice. One of the very few occasions in which I experienced carburetor ice was at 7,500 feet on a clear summer day. In a Cessna 170 my engine gradually began to lose power and to show signs of roughness, symptoms of carb ice, but because I was flying in blue skies and sunshine, I just could not believe that what was happening was the gradual accretion of ice in the carb throat. (Remember our discussion of "denial" in Chapter 4?) However, when I applied carb heat, the engine coughed a couple of times and cleared right up.

The only other time I experienced carb ice dispelled another myth for me. I had always been taught that in a full-power climb, it is very unlikely that ice will form in the carburetor. Not so. On the climb out in a Cessna 150 in a very light rain shower, the symptoms appeared and again went away with the application of carb heat. Other than those two occasions, both at times when I least expected it, in over 40,000 hours of flying I have never experienced induction icing. This is probably because as soon as I enter a cloud, any cloud, when in an airplane with a carbureted engine, the heat goes on. Not only do I apply carb heat, but I also apply pitot heat prior to entering IMC. I once had an airplane with a good working Pitot heat system, but every time I flew the airplane in any icing conditions (it was certified into "known ice"), I would lose my airspeed indicator, so please understand that this can happen even with the application of pitot heat. Of course, losing your airspeed indicator is no big deal. Remember, if your power setting is right, and the airplane attitude is right, the airspeed simply has to be in the ballpark! (See Chapter 2.) And if you are thoroughly familiar with the airplane you're flying, you know just what power settings will produce which specific results in terms of performance. Remember also our friendly formula that always works in any flight situation: attitude plus power equals performance.

Although with a carbureted engine, as I said, I apply heat as a preventive measure before I really need it, when flying an airplane with a fuel-injected engine (or engines), I do not go to the alternate air intake until I actually observe a power loss. In either case, handling induction ice is a fairly simple matter. Get air for the engine from a different source; in the one case from the carburetor heat exchange box, and in the other from inside the cowling where the air is warmer than the cold, moist ambient air that is impacting the outside air intake. If ice builds up in the induction system of the airplane, the engine or engines just won't do their thing. Remember, an internal-combustion engine requires just three things to keep it going; fuel, spark, and oxygen (air), and if any one of them is not present, your airplane becomes a glider, and a very inefficient glider at that.

Structural ice presents an entirely different problem for the aviator. When I stated that ice is really weird stuff I wasn't kidding. I meant exactly what I said. I remember one time while flying along in IMC I picked up about a quarter inch of rime ice on the leading edges of the wings and tail surfaces (and the inboard portions of the prop). With absolutely no apparent change in conditions, it just quit growing. I mean the temperature and moisture were exactly the same and I got no more ice. It didn't go away, but it didn't get any worse, either. Weird? I'll say! To this day I've never been able to figure that one out.

I don't care how much de-ice and anti-ice equipment you have, certified into known ice or not, the thing to do when you see ice, any ice growing on your airplane, is go away! If you avoid ice just as you do thunderstorms, it can't bring you down. With respect to which way to go when you go away, it's up to you. My own policy is as follows: unless I know with absolute certainty that there is warmer air below, as soon as I see ice starting to form anywhere on the structure, I start screaming for higher. This policy is based on the knowledge that you can always come down, but unless you do it quickly, you may not be able to climb. If you can get above the cloud deck and into clear air, the ice will sublimate away. Some experienced pilots take the opposite position and start to descend when they blunder into icing conditions. I fail to understand the logic in that approach, but maybe those pilots know something I don't.

Please understand: the sublimation process is agonizingly slow. I remember a specific trip in a Cherokee 180 in which I picked up a quarter inch of rime ice on the climb out. I leveled off at 7,000 feet in blue skies and sunshine above a solid deck, which was perhaps as much as 1,000 feet thick and which started about 1,500 feet above the surface. After I had flown approximately 100 miles I couldn't tell if there was any less ice than had grown on the wings during the climb through the clouds. Another hundred miles along the way and there was perceptibly less ice, but barely so. Two-hundred miles farther, as I approached my destination, I noticed it was all gone. This after well over an hour of flying in clear air. Of course, I picked up another quarter inch on the descent. This is how slowly sublimation works. Also, by climbing you might be able to get between layers, and although you may not get rid of any ice you've acquired, you're not likely to pick up any more.

On the other hand, as opposed to the slow sublimation process, if you can find warmer air (above freezing temperatures), any ice you have picked up will come sliding off in sheets. On this subject, you should know that although any ice is bad, clear ice is much worse than rime ice. Rime ice gets its white, milky appearance from the fact that it is full of tiny air bubbles and therefore it weighs much less than a comparable amount of clear ice which is very solid and quite heavy. Also, rime ice builds forward from the leading edges of whatever surfaces it can find, while clear ice spreads over the entire upper surface of the airplane (and wings) and tends to destroy the shape of the airfoil from which we get our lift.

What do we know about clear ice that's good? Very little, but I'll tell you this: it usually comes from freezing rain. What does this tell us?

It tells us that there is warmer air above. To get freezing rain there must be an inversion! The moisture starts out as rain and as it falls through colder air, it becomes supercooled. In fact, it gets below freezing and should freeze into ice. But it continues to fall while it is still water because its so dumb it doesn't know enough to freeze. However, when it strikes a surface, splat, instant ice. Clear ice, as mentioned, builds amazingly fast. In fact, it has been known to come on so fast that no amount of ice protection can get rid of it as fast as it is forming. I know of at least one corporate jet that was brought down by rapidly forming clear ice in spite of having the very best anti-ice and de-ice equipment that was then available.

Now this inversion from which freezing rain is coming is usually 1,000 or less feet above where we're flying, so if we have the capability to climb, we may be able to get up to where the temperature is above freezing, and the airplane will shed the ice almost as fast as it picked it up, perhaps even faster. When you lose ice this way, it comes sliding off the wings in large sheets. These are the reasons I try to climb rather than descend as soon as I start picking up any ice.

There are those who favor starting to descend if and when they start to observe the accretion of ice on their airplanes. I have a policy of never telling anybody how to fly an airplane. Everybody should do that which is most comfortable for him or her. However, what I do is tell people what I do and why. If they like what they hear, they can do as I do; if not, they can do as they please. Nevertheless I don't think it is wise to start to descend as soon as the ice starts to form. You can always come down, but you may not be able to climb.

Of course, the best way to deal with ice is to avoid it entirely. If you don't go where ice is, it can't get you in trouble, but if you fly IFR in the northern United States in the winter, sooner or later you are bound to encounter icing conditions. When you do, get out! Go someplace else. And do it now! I mean immediately, right away. Don't hang around to see what comes next because I guarantee you won't like what you see.

# II Thunderstorms

Back when I was trained and when I first started giving instruction, there was no instrument training at all in the primary phase of flight training. In fact, other than the engine gages and a wet (magnetic) whisky compass, there were no instruments whatever in the airplanes, and no radios either. I'm not sure this was such a bad idea. The students were so scared of being in IMC, they never became confident of their ability to do so, as some of them now do.

We scared our students by telling them that if they flew into a cloud, they'd die. It was a simple as that. "See that cloud over there? If you go in there, you will die!" The pilot with an instrument rating was a rare individual indeed. Of course, instrument flying was a great deal more difficult without VORs, Loran C, GPS, DME, and all that fancy stuff we have today. Consequently, other than the air carriers, nobody flew when the birds were walking. When the weather was IFR, most pilots stood down. If they encountered weather en route, they landed and waited it out. I suppose it was good for the character. Among other things it forced you to learn to be patient.

Today, however, the situation has changed. There are great numbers of people flying around in cloud under IFR, and not only must they develop new skills, but they must acquire quite a bit of additional knowledge as well. I have formulated a personal rule based on a single terrifying experience. As a result of having been vectored right into a tornado, I flat-out refuse to fly anytime there are embedded cells in the area and I do not have airborne weather avoidance equipment (Radar, Stormscope, or Strikefinder) aboard. The power of nature is awesome indeed. Everyone has seen pictures of what really severe weather can do, and its destructive force is absolutely terrifying. I have had successful experiences with ground-based radar, but I won't count on it again. One time when I was flying from New Orleans to Detroit City, there was a front lying diagonally across my path with thunderstorms all along it. As I progressed, I kept deviating farther and farther to the right until I found myself heading southeast. Finally, the controller at Atlanta Center said, "If you want to get to Detroit, sooner or later you are going to have to turn North."

I replied, "Find me a soft spot, and I'll be happy to turn North."

He then gave me several vectors that put me through just such a soft spot as I had requested, and I made my way to the destination. But because of another experience I had, I will never again depend on ground-based radar to tell me where to go when there is heavy weather in the area.

At a time when my company had flight schools at three different airports, I was commuting daily between two of them while my associate acted as manager of the third. I personally supervised the operations at both the Detroit City Airport and the Oakland-Pontiac Airport, while Don ran the operation at Flint-Bishop Airport. My home was closer to the Pontiac Airport, so my policy was to start the day there, put in a few hours, take one of the school airplanes (we had 19 in the fleet), fly to Detroit City, work about half a day, return to Pontiac, and put in a few more hours prior to going home. Made for a long day, but it was fun.

A few years ago on a quiet Sunday in April, after spending the usual 3 or 4 hours at Pontiac, I took our Aztec (N4647P, which would ever after be known as Forty-seven Door-popper) and flew down to Detroit City. One of our customers, a private pilot, came along for the ride. The weather was not conducive for flight training, so after about an hour, I gathered up the weekend receipts and headed back.

Since Don had taken one of the two place trainers from Pontiac to Flint and was scheduled for a 6-month check in the Aztec the next day, I thought it might be a good idea to take the Aztec to Flint for him to get some practice in on the way home. I would return from Flint to Pontiac in whichever trainer he had flown up in that morning.

The weather en route and at both ends was reported to be 2200' broken, 3500' overcast with 4 miles visibility in light rain and drizzle, so I departed VFR, planning to fly at 2700' MSL (the surface in the area is between about 700 and 950 MSL), and the MVA (Minimum Vectoring Altitude) is 2700.

After liftoff from runway 33 at Detroit City, I climbed straight out (right on course to Flint), and leveled off at 2700. I called Detroit Metro Approach (the radar facility for the area) and requested VFR advisories en route to Flint. The controller responded, "Radar contact." And he gave me a squawk code. After a very few minutes, I noticed wisps of cloud forming beneath me, so I called Metro and advised the controller that we were rated and equipped (I was instrument rated and the airplane was equipped for instrument flight), that I would soon be unable to maintain VFR, and I requested an IFR clearance, "present position direct Flint."

I was cleared as requested, given a new squawk code and told to climb and maintain 3000. As I reached 3000, I was solid on the gages, and I began to encounter turbulence. I don't mean light chop, or even moderate turbulence. I was getting the kind of jolts in which you bump your head on the ceiling (hard) even with the seat belt cinched up tight. I asked the controller if he had his scope on circular polarization, which wipes out the weather, or was he painting any weather and he replied, "Both!" This was my clue that it was indeed heavy turbulence, as if I didn't already know. If it is so bad that it comes through even when he's got it turned off, it has to be really bad. It also told me that I ought to be somewhere else. I thought of my dry house and warm wife, and the fact that Don wasn't even expecting me in Flint. I, therefore, called the controller and advised him that I wanted to amend my destination. I would go to Pontiac instead of Flint.

My location at that time was about 8 miles Northeast of the Pontiac Airport, halfway between Detroit City and Flint. (See Fig. 9-1.) At that time Pontiac had two IFR approaches, one for each end of the 5,000-plus-foot east-west runway. Both were VOR approaches. The controller told me to turn right (away from the airport), heading 180 to permit me to intercept the final approach course for the VOR 27 approach outside the final fix (Keego Intersection). (See Fig. 9-2.) As I started the turn, all hell broke loose. It was just like driving into a brick wall.

The door was ripped from the aircraft, leaving a circle of jagged aluminum around the receptacle where the upper door pin is supposed to secure the door and blank spaces where the hinges used to be! (See Fig. 9-3.) The airplane was pitched inverted, plastering the microphone, which I had released, against the headliner. We were being extremely violently tossed about. Because of the unbelievable pressure changes, the airspeed indicator was rapidly going from 0 to 200 miles per hour, and the altimeter was showing 1,000 feet and spinning around to 6,000 in nothing flat. The airplane was equipped with an instantaneous vertical speed indicator, and it was registering a climb of 2,000 feet per minute up and within a matter of a second or so, 2,000 feet per minute down. About then the baggage door departed the airplane as well.

I had no idea what our attitude was. All the gyroscopic instruments had tumbled, and the magnetic compass was bobbing around like a cork in the ocean in a storm at sea. More by instinct than thought process, I retarded the throttles, and since I was being shoved against the belt to my left, I stomped on the left rudder. The first

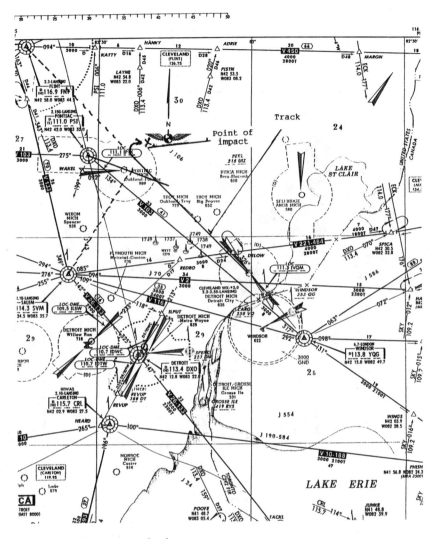

**9-1** *VOR 27 approach chart.*

thought that flashed through my mind was, "That damned door has plunged through the roof of a house and killed a baby sleeping in a crib." I next came close to allowing the infamous hazardous attitude of resignation to take over. I was thinking that after 35 years and well over 20,000 hours of flying, I was finally about to buy the farm. This is it. That's all she wrote. These two thoughts must have flashed by incredibly fast because I then recall telling myself, "Okay, you dummy, if you settle down and do what you know how, you just might survive."

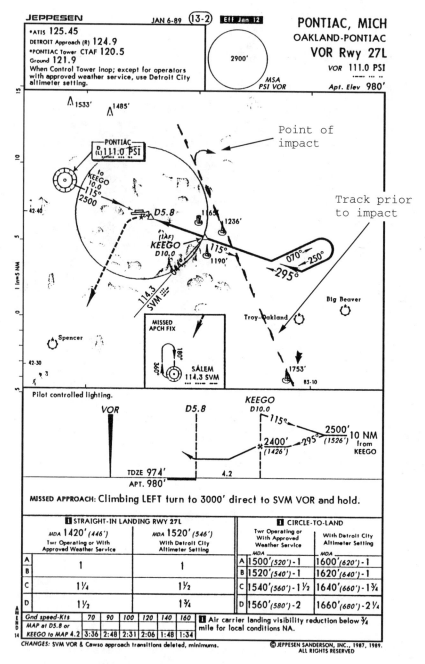

9-2  VOR 27 approach chart.

**9-3** *Photo taken immediately after I landed. Notice the coating of ice on the airframe and how the top pin and the hinge were ripped from the door frame.*

While I was struggling to regain some semblance of control over the airplane, the approach controller was reading me Pontiac weather, which he said was 1000 and 3. I grabbed the mic and advised him of the fact that our door was off, not just open as sometimes happens. He asked if I wished to declare an emergency. Now, I'm not one of those pilots who are concerned about the potential paperwork associated with declaring an emergency, but I do know that the controller's manual by which they have to live requires that they illicit a bunch of information when a pilot declares an emergency, including the amount of fuel aboard, number of passengers, etc., and since I had my hands full of an airplane and no time to bother with answering that stuff, I replied, "Negative. Cancel IFR." I was still IMC, but I couldn't be concerned with further communications at that time.

With the power reduced, I applied approach flaps. I was still holding full left rudder to keep the wings what I believed to be level. When the nice man said that the ceiling at Pontiac was 1,000 feet, I gained a little confidence, for Pontiac's field elevation is just under

1,000 feet, so I thought I'd break out about 1,000 feet above the ground. No such luck. I continued to descend until I made ground contact, at which time I was only a couple of hundred feet above the surface. Now, with a horizon, I could at least keep the damn thing level, although I was still having a great deal of control difficulty.

I looked around and spotted a large water tank. Believing it to be the one located 5 miles Southeast of the Pontiac Airport, I started flying Northwest, orienting myself by the section lines on the ground. (The whiskey compass was still useless and wobbling so as to be unreadable, much less reliable, and who knows where the directional gyro or heading indicator had ended up? Remember, they had both tumbled.) I switched over to the tower and reported, "Waterford tank inbound."

The local (tower) controller advised me of the wind (300 degrees at some value or other), asked me if I wanted to declare an emergency (he had been advised of our plight by the approach controller at Metro), and cleared me to land on runway two-seven. I again turned down the offer of declaring an emergency (this time because I thought I knew where I was), and I kept going. Long after I should have reached the airport (must have been 5 or 6 minutes), I spotted a transmitter antennae tower straight ahead. It's top was in the cloud, higher than I was! I did a quick 180. I then knew where I really was, about a dozen miles north of the airport. I called the tower and reported that I was now north and headed back toward the field (see Fig. 9-2).

When I was just off the Northeast boundary, the local controller advised me, "We've had a wind shift. The wind is now 360 degrees at 38 knots, gusting in the forties. Thirty-six is favored, but you may have 27 if you like."

I looked out that great big 4-foot hole in the side of the airplane where there used to be a door, and there was that gorgeous runway 27, over 5,000 feet long and 300 feet wide. Thirty-six is the shortest runway at the airport, being only 1,843 feet long and 48 feet wide, but it has a long overrun at the north end before you run into the fence. I opted to land into the wind on 36.

I extended the landing gear and flaps, made a right 180 to final and executed one of the best, smoothest landings of my long career. I had retarded the power on short final, and just as I started to flare, I added a bit of throttle. The left engine responded and the right engine ground to a halt. I don't know just when I lost my right engine, but I'm sure that loss contributed to my control problems. (It was due to ice in the carburetor.)

The tower controller said I could make a 180 on the runway and taxi back to my office, "and by the way, it appears that a piece of that door is stuck in your tail." I told him that if it was all the same to him, I'd just sit where I was for a few minutes while my heart stopped racing and my breathing returned to normal. My passenger leaned out the opening on his right and looked back toward the tail, and he fainted! Throughout the entire episode he had sat terrified in the right seat with both hands holding his seat belt with a death grip, saying over and over, "I'm glad you're flying. I'm glad you're flying." I don't know if this was because he thinks I'm a skillful pilot (mistaken idea if that's what he thought), or merely because he was glad it wasn't him.

When I got the airplane parked and we deplaned, I looked at the empennage, and I, too, almost fainted. The entire vertical stabilizer forward of the main spar was totally demolished, and the mangled remains of the door (all of it) was stuck in the tail. The baggage door was never found. The airplane had approximately $\frac{1}{8}$- to $\frac{1}{4}$-inch of ice all over it. (See Fig. 3-1.)

I went up to the tower to file a report and thank both the local controller and the approach controller, and I listened to the tapes of my communications with both. I'm proud to say my voice was perfectly calm throughout. After the initial moment of terror, I had applied my training and experience to survive. I was also lucky. About 10 minutes after I landed, Pontiac went to zero/zero! The approach controller told me that when he turned off his circular polarization, the echo where I was had a hook at the bottom end. This means he had vectored me right into a tornado!

The following day a general-aviation operations inspector from our local FSDO sarcastically remarked, "You no doubt failed to latch the door securely." Obviously I had done so, as the photo shows. The top pin and the hinge were ripped off by the force exerted by the pressure differential.

An airworthiness (maintenance) inspector from the FSDO also had a remark to make, and this one broke my wife up. (She happened to be in my office at the airport at the time.) The guy picked up the crumpled piece of aluminum which used to be a door and with a perfectly straight face said, "I don't think I'd bother trying to repair this."

Three days later a young lady whom I had never met, and whom I haven't seen since, stopped by my office at the Pontiac Airport and graciously presented me with three photos her husband had taken right after I parked the airplane. It seems that her husband was a free-lance flight instructor and they lived a few blocks from the airport. He had been listening to his air-band radio and heard the entire episode. As soon as he knew I was safely on the ground, he rushed over and took three photographs of the airplane with the door stuck in the tail. Or perhaps he was there waiting to see and photograph the crash? (Fig. 9-3.)

None of the terrible weather that descended on the area that day was forecast. It was all unexpected. The lesson to be learned from this adventure is to keep your cool, apply your training and experience, and there's almost nothing that can happen in light aircraft that isn't survivable.

Personally, I have never since that day been in clouds where there is the potential for embedded thunderstorms without some sort of weather-avoidance equipment aboard. I mean on-board weather radar, Strikefinder, or Stormscope. The ideal is to have both radar and either of the others so you can interpolate the information from all sources, but if I had to choose, I'd take the Strikefinder or Stormscope instead of radar. After all, radar tells us where the precip is, and the heaviest turbulence is rarely colocated with the heaviest rainfall—it's off to the side a few miles away and may not show up on radar as a particularly bad area. I'm willing to fly through all the water the sky can put out as long as I'm in smooth air. What the Stormscope and Strikefinder show us is where the chop is, and that's what I want to know—and avoid. (See Fig. 9-4).

# III Fog

Fog, which severely limits visibility, can be among the worst weather hazards a pilot may encounter. Here in the Great Lakes area where I live, every fall (usually about mid-October) and every spring (usually in March) we get a period of four or five consecutive days when we get to hear that rare term, "zero-zero" at many of the local airports. We get this in the morning, starting about sunup and lasting until almost noon, by which time it has burned off.

In the discussion of icing above, I mentioned the fact that the instrument pilot becomes aware of and pays close attention to that

**9-4** *Another view of 47 door-popper with the door stuck in the vertical stabilizer.*

little round dial which tells him or her what the outside air temperature is. All pilots also pay close attention to the temperature-dewpoint spread at the surface, or if he or she doesn't, he/she certainly should because it is literally impossible to land if there is fog laying on the ground at your destination. Bob Buck, in *Weather Flying*, has this to say about dewpoint:

> *There are a couple of items that pilots should know. One is dewpoint. Most of us know it as the temperature at which condensation begins. If the temperature is 50 degrees and the dewpoint 45 degrees, we have only to cool the air 5 degrees for the moisture to come out where we can see it...and if it's fog, that's all we can see...*
>
> *Sometimes fog doesn't form until the sun comes up. An airport may have the same dewpoint and temperature all during a still night and yet not fog in. Then, just as the sun is coming up and we think everything will be okay, the airport goes zero-zero in fog.*

In discussing the weather requirement for filing an alternate with instrument applicants, I always point out that there will be occasions (although rarely) that a given flight will be impossible because of the fact that there will be no legal alternate within the range of the airplane. This is the one situation where the air carriers have the advantage over those of us who fly general aviation airplanes. I fly out of Pontiac, Michigan, and there have been times when I have filed for Chicago and had to list Pittsburgh as my alternate. But there are occasions where the entire eastern half of the country is socked in, from St. Louis to the Atlantic Coast. When this situation prevails, we are simply stuck, but United Air Lines can file from Detroit to Chicago and list Los Angeles as their alternate.

Like ice, fog is funny stuff. I remember one time holding over the outer marker in absolutely clear skies with great visibility when my destination airport, about 5 miles away, was covered with ground fog about 100 or 200 feet thick. I was behind schedule for picking up a passenger, but I held for about a half hour, waiting for the fog to burn off. Finally, I could see the runway, and I started down on the approach. I landed at the approach end of the runway, and on the rollout about a quarter of the way down the runway I ran into the fog. It was just as if a curtain had been drawn across the runway at that point. I couldn't even see the side edges of the 300-foot-wide runway, of which I occupied the middle (I was taught to always try to land on the centerline, except under very special circumstances). Again, in *Weather Flying*, Bob Buck puts it this way:

> *We can fly in clear weather, day or night, and note that our destination reports zero in ground fog. We get over the field and are surprised to see the runway and airport below us well in view.*

> *This is simply because we are looking down through a very shallow layer of fog, rather than through it horizontally. But if we say, "Heck, I can land in that," and make an approach, we'll get an awful shock as we descend and suddenly, about the time we start to flare, go on solid instruments.*

When I arrive at my destination and find the weather to be right around minimums, or even reported to be a bit below minimums for landing, my policy is to go ahead and execute the approach, expecting that I will probably have to miss. But if there is fog on the ground, I don't even go down for a look-see. I know I won't make it in for a successful landing, so there is no use trying.

# IV  The best weather information

The very best weather information a pilot can get is by talking to the pilot who was just there. As you're charging down the airway, either talk directly to the pilot you're following, or ask ATC to ask the pilot how his or her ride is. This is the best way to get really current and valid information. After all, the meteorologists and controllers who are getting their weather from a scope are no doubt in a room with no windows and may not have any idea what's really going on outside. However, the pilot who is flying up ahead of you on the airway is experiencing what you will shortly be getting.

When you land for fuel (or other purposes) and intend to take right off and continue your journey, try to talk to a pilot who just came in from the direction you're going. This is another source of valid and current information. As I'm sure you know, on the whole pilots are prone to be extremely helpful to their fellow aviators. And if you can talk with a pro, a cargo hauler or charter pilot, all the better, but discussing the weather up ahead with a pilot who has just been there tells us what we need to know.

Of course, nothing can replace your own eyeballs. If there are isolated buildups in the area and you can see them, you can certainly circumnavigate them. How far away should you stay to avoid the most severe chop? That's up to you, but I like to give even small cells at least 15 miles. Even if there is a fairly solid-line squall and you're on top of the solid deck, you can see the columns shooting up and weave your way through them, you can "ride the saddlebacks." You can sometimes even do this when you're between layers if the outside visibility is good. On occasion I have had to penetrate a solid line, and with help from ATC controllers with Radar, I was able to find soft spots and work my way through, but I don't recommend this as a regular policy. The thing to do is avoid that kind of weather altogether. And with respect to the "classic 180," unless you execute this maneuver before you enter a cell, it is probably better to go straight ahead. You'll likely pop out the other side before you would get out by going back, and you won't risk losing it in the turn. I know this from experience.

I certainly don't ignore forecasts, but I am skeptical of them. I never tell anyone how to fly. What I do is explain what I do in a given situation, and why. Then it is up to each individual to accept or reject my policy. Each pilot must do what is comfortable for him or her. My own policy with respect to launching into what has been forecast as particularly nasty weather is to go ahead and take a look. I can always turn around and run back from whence I came. I often find that I can work my way through or around whatever is out there. I have thus completed many a trip that would have been canceled if I had relied totally on the forecast weather. Of course, I have, on occasion, turned around and returned to my starting point. But whenever I did so, it was with the knowledge that I had been confronted with an impossible situation. You must do what works for you.

I apply the same principle to approaches. Back when the reported visibility was the key to open the door to the approach at the initial fix, I diverted to some place other than my intended destination a lot more than I do now. Ever since flight visibility became the determining factor, I go ahead and take a look. More often than not, when I arrive at the minimums for the published approach I have the runway environment in sight and am in a position from which I can make a normal landing. Also, since I'm certainly not the world's greatest pilot, but just a tired old man and an average pilot, when the weather is at minimums if I miss on the first attempt, I usually will take another shot at it, on the theory that I probably didn't do something right the first time. Then, if I don't get in, I know it is time to go someplace else and land (see Chapter 5, confession time), although on occasion I have held in clear air at the initial approach fix waiting for the fog to lift off my destination airport.

Again, I'm not telling you to adapt my policy to your own flying. I'm merely telling you what I do in the situation. It works for me and has permitted me to complete more flights as planned than I would otherwise have been able to do. If you think this procedure has merit, go ahead and try it. Unless you're picking up ice or have to penetrate a cell to get there, in which case you shouldn't be there in the first place, you have nothing to lose by taking a look-see. If you don't like what you see, you can always go somewhere else. That's why we carry a good fuel reserve.

# V PIREPS

Under some conditions, workload permitting, whichever ATC facility you are in contact with will solicit PIREPS (pilot reports on inflight weather and flight conditions), but you certainly don't have to wait to be asked. This is one situation when my personal "never volunteer" rule does not apply. And if you, too, have such a rule for yourself, I urge you to not apply it to giving PIREPS. Be generous with information that may help your fellow pilots.

Anytime you are on a trip lasting more than 30 minutes, it is a good idea to check the current weather at your destination by calling Flight Watch (EFAS) on 122.0, and while you are at it, why not forward a PIREP? Even if the weather is good, you can tell 'em that. You should include your location, type of aircraft, altitude, and comments. It makes no difference what sequence you relay this information in. The person taking it down will put it in the proper sequence, and if you left anything out, the person will ask and then forward the information to the Flight Service System for dissemination.

If you are busy talking with Center, give them the PIREP and they will forward it along. In other words, it really doesn't matter which facility you give it to because it will wind up where it belongs. And when doing your preflight planning, be sure to ask for a standard briefing and have the patience to listen while the briefer goes through it all, including all the PIREPS and NOTAMS that will affect your flight. PIREPS come in two forms, UA (routine) and UUA (urgent) and are transmitted as such. Pay close attention to both PIREPS and NOTAMS because they are conveying information you need to know, or at least stuff that is bound to prove helpful. And remember, ignorance is no excuse if you should violate a regulation simply because you didn't have all the information available. As pilot in command you are required to obtain all pertinent information prior to embarking on any flight.

I feel that since I take advantage of the PIREPS forwarded by other pilots, I have a duty to forward the information on what I am experiencing, don't you?

# 10

# Flying different kinds of equipment

## I The taildragger

Flying an airplane with a tailwheel (I still call 'em conventional-gear airplanes), is quite a bit different from operating one with a training wheel out in front. To begin with, there's the matter of center of gravity location. In a trigear airplane, the CG is forward of where the pilot is sitting, while in a taildragger it is aft of the pilot's location. This fact introduces a much greater potential for groundlooping than is the case with a trigear airplane. In a stiff crosswind while on the ground, if the pilot isn't sharp and prompt in acting to control his airplane, the tail will start to swing sideways, and in nothing flat the airplane will swap ends with itself. Simultaneously, a wing will drop, possibly far enough to encounter Mother Earth, resulting in damage to the machinery and embarrassment to the pilot, and in extreme cases the airplane will flip clear over, resulting in severe damage and possibly injury to the human being or beings aboard.

Since shortly after the Wright brothers proved that heavier-than-air flight is possible, the controversy has raged as to which is better, full stall (three-point) or wheel landings. Like most other things which are matters of technique, they both have merit, and both should be mastered for use when the situation is appropriate. In other words, neither is superior to the other; they are simply different. It just depends on the circumstances and the specific make and model of airplane involved. Some airplanes seem to prefer the full stall over the wheel landing, while the reverse is true of others. When I was trained, and when I instruct in a taildragger, I was taught and I teach the three-point landing first, and when the student masters that, we move on to wheel landings. The wheel landing does require a some-what longer runway because the airplane is going forward faster at the moment of touchdown, but on the other hand, the wheel land-ing makes the airplane much more controllable in a stiff crosswind.

Usually the student learns to make three-point landings first, and then advances to the wheel landing, probably because it is easier to go that way. With the correct "sight picture" (see Chapter 2), landing attitude stalls can be practiced at altitude and then brought down to a couple of inches off the ground where the airplane is held off as long as possible. Finally, when it quits flying, if the pilot is prompt and effective with rudder control, the airplane will be rolling straight ahead on the ground. The stick or yoke must be held full back as soon as the airplane is on the ground to prevent the possibility of it taking off again. But if there is a moderate to strong crosswind, there is more likelihood of a groundloop with the full-stall landing.

The wheel landing is somewhat more demanding, and because of the excess forward speed at the point of touchdown, it requires a somewhat longer runway. Some things just don't seem right, and high on the list is using forward elevator in a taildragger on the ground, a technique required in making wheel landings. You practi-cally have to force yourself to do it at first, but once you become convinced that the airplane won't nose over, it doesn't seem so bad. As soon as the main wheels kiss the ground, with the airplane in a level attitude, the stick or yoke must be pushed forward to hold the airplane on the ground, and when the speed dissipates to the point that the tail drops by itself, the stick is held all the way back as in a three-point landing. You certainly don't want to bring it back too soon (during the landing roll with the tail off the ground for in-stance), or you'll find yourself flying again.

It is recommended that for both short- and soft-field landings, the full-stall method be used because of the slower touchdown speed. The object in both cases is, of course, to be going forward as slowly as you possibly can at the moment of touchdown. If there is a strong crosswind (requiring a wheel landing) and a short runway (requiring a three-point landing), your best option is to go someplace else, land, and wait for the wind to swing around to a more favorable di-rection or to subside. For a full discussion of taildragger operations, I refer you to Harvey S. Ploudre's excellent book, *The Compleat Tail-dragger Pilot*. It is a comprehensive manual for all who fly taildrag-gers, and I believe it to be the definitive work on the subject.

# II Multiengine flying

As a general rule, the larger and heavier the airplane is, the easier it is to fly in terms of herding it around the sky. The only thing about twin-engine airplanes is that there are two of everything associated with the power plants. Otherwise, the twin is just another airplane. Flying a twin is not particularly difficult. They all respond to the same immutable laws of nature, but there are two things the multiengine pilot must do. First he/she must thoroughly know and understand the systems in the airplane, and second, he or she must know and apply the procedures. It is absolutely essential that airspeeds be flown precisely. If you don't know the systems in your multiengine airplane, and you fail to perform the appropriate procedures, the penalty is very severe. You will die in a twin-engine airplane.

Statistics indicate that per amount of exposure, there are more fatalities in light twins than in single-engine airplanes. So much for owning and operating a twin because you think it is safer—after all, you've got a spare engine, don't you? There are numerous theories advanced as to just why more people die in twins than in singles based on the number of hours flown. Personally, I think there are three basic reasons for this. First, the people who can afford to own and operate multiengine airplanes are likely to be high-powered business or professional people, the kind of people who are used to making quick decisions in their everyday lives and who fly, at least in part, for relaxation. Since aviation for them is, among other things, a means of escape and relaxation, they are apt to become complacent and careless. They are also likely to be the kind of individuals who are prone to overconfidence.

The second theory which I believe plays a part in the rather negative accident statistics with respect to twin-engine flying is the temptation offered by all that redundancy. If one taxis out to depart in a single-engine airplane and on the pretakeoff engine runup gets a 600-rpm mag drop, the pilot turns around, taxis back, and turns the airplane over to a mechanic to fix! The same pilot encountering the same situation in a twin is likely to say, "What the heck! I've got three more magnetos." And the pilot goes ahead and launches, even though there is a known deficiency. Enough exposure to this kind of thinking is guaranteed to get a person in trouble.

But the most important theory of all, in my opinion, goes like this: during many-motor training, after the first few sessions during which the students learn where all the knobs and dials are located, they never see a time when everything is all working at once. It is just one emergency after another until the students become accurate, smooth, and rapid in their reactions to emergencies. (And on the subject of rapid, nothing says one cannot be both smooth and rapid.) Then the students take their checkrides and acquire the rating. For the next 6 years or so, they fly from point A to point B and everything works just fine. Then one day it doesn't, and they're not ready for it. The cure for this is quite simple. The conscientious multiengine pilots see to it that they get recurrent training regularly and frequently. This is the best insurance a person can have. If you continue to expect the unexpected and are prepared for whatever might happen, you're almost certain to come out okay. The objective is to avoid being caught by surprise. If there are no unpleasant surprises, multiengine pilots are likely to indeed enjoy safer flying than their single-engine counterparts. And remember, nothing is happening so fast that you have to panic. It's panic that's the killer! Remember our discussion of mental attitudes in Chapter 4 and the thinking process in Chapter 5!

I scrupulously avoid absolutes such as "never" and "always," but I can honestly say I have never taken the runway to depart in a twin-engine airplane without having picked out a point somewhere down the runway by which I am going to be off and flying, or I'm simply not going. This system has forced me to abort a takeoff or two, but I'm still here. Please don't misunderstand. I don't always compute an accelerate-stop distance, but I do have a spot picked out. *Accelerate-stop* is defined as the distance in feet required to accelerate to liftoff speed, have a catastrophic engine failure, allow a brief period for reaction, and get the airplane stopped while you're still on the pavement. Under Part 91 of the regulations, there is no requirement that the pilot of a light twin have an accelerate-stop distance for takeoff planning. This means there are likely to be occasions when the operator of a twin-engine airplane will land at airports on which the runway is not long enough for an accelerate-stop distance when the pilot is ready to depart. In those cases, pilots must treat the situation as if they were in a single-engine airplane, and in case of an engine loss just after liftoff, they must land more or less straight ahead, no matter what lies in the way. Just pick the softest area. (See Chapter 3 on the subject of crash survivability.)

With the dependable engines we have today, a good, high-performance single will do virtually anything for the pilot that a twin will, but for hard IFR, over water, and night flying, most of us feel more comfortable in a twin in spite of the dependability of today's engines. And if multiengine flying is approached properly with regular and frequent recurrent training, there is no doubt that it can be safer than tooling around in a single. For example, I flat-out refuse to cross Lake Michigan in a single-engine airplane. The bottom of that lake is literally paved with dead airplanes, and I have no desire to join them. (Note: I do cross Lake Erie in a single if I have everything stabilized and am at 9,000 feet before I start across. There is a string of islands all the way across, and an airport on every one of 'em.)

# III Soaring

Although all kinds of sport flying is for fun rather than transportation, all you power pilots don't know what fun is until you try soaring; it's so quiet the instructor can hear the students cry. Soaring in a glider or sailplane is just pure joy. With today's sophisticated aircraft that most of us fly—even the basic four-place fixed-gear, fixed-pitch prop airplanes—the pilot is basically an equipment manager rather than an airplane manipulator. Flying a glider, on the other hand, is pure stick-and-rudder flying, which is based on the skill of the pilot rather than the sophistication of the equipment, and believe me, it is pure fun! In fact, it has often been said to be the most fun you can have with your pants on. It has been accurately said that soaring is to power flying as sailing is to power boating. Or expressed another way, soaring is really three-dimensional sailing.

Robert N. Buck is a retired airline captain who flew everything from DC-2s to 747s, and in his excellent book, *Flying Know-How*, he says this about flying gliders:

> *I fly gliders for fun.... and because its the most fascinating flying there is.... to me.*

> *I'm not trying to sell gliding, or soaring as they call it. I don't manufacture gliders, distribute, or sell 'em. I don't own stock in a glider factory, and actually it's to my personal, selfish advantage if the world has fewer gliders; then I won't have to wait for tows on busy days or fly against so many hotshots in contests.*

*But I've found soaring makes good pilots, pilots who understand weather and the techniques of flight better than those who don't soar....*

*Soaring means learning to fly on the ragged edge of stall because much of glider flying is turning in thermals to climb, and you want to be on the very slow side to keep the turn radius small and be at the best speed for maximum lift, which isn't far above stall. So flight near the stall becomes routine.*

For years I had ridiculed the idea of flying anything without an engine to keep it up, but I learned better the first time I went up in a glider. I fell in love all over again. I had been occupying various portions of the airspace for over 30 years when I was inspired to try soaring. My wife and I were in a hotel in Innsbruck, Austria, waiting for breakfast early one beautiful morning when I looked out the window of the hotel room and saw two sailplanes going back and forth along a ridge which rose straight up for several hundred feet above a meadow on a plateau in the Alps, just outside the hotel window. They were so graceful as they flew silently by that I was inspired to say to my wife, "Just as soon as we get back home, I'm going to undertake glider training." And within a year I was not only a certificated glider pilot, but a glider instructor and designated glider examiner as well!

Most glider training in the part of the country where I live is accomplished through soaring clubs, but there is an excellent commercial glider school about 120 miles from where I'm based, so as soon as we returned from Europe I rushed right out and signed up for a commercial glider transition course. This being October, the glider school had cut back their operations to weekends only. They operate daily through September, but in the fall the regular soaring season ends and they revert to weekend-only operations.

However, the school operator agreed to have a tow plane and pilot and a glider instructor meet me each Wednesday afternoon for the next three weeks so that I might complete the transition course. At that time the experience requirement for transition from power to glider was the same for private or commercial privileges. It amounted to only 10 solo flights, during each of which at least one 360-degree turn was required. The only real difference between the private and the commercial was in the skill requirement. For the commercial, substantially more precision was required. For example, the spot landing allowed much less tolerance for commercial privileges than was required for private privileges.

I had bought an 18-flight package, and I had a very good young lady instructor who was an excellent teacher and who soloed me on my second tow. When I added the words "Commercial Privileges—Glider" to my pilot certificate, I still had two flights left on the 18-flight package I had purchased, so I returned the following Sunday and gave my wife and my sister each a glider ride. This particular glider school uses aero tows to launch the glider, so my glider rating was restricted to "aero tow only."

During that winter I took the written examination for adding "glider" to my Flight Instructor Certificate, and the following summer I spent every Sunday afternoon training for the glider instructor rating at a local glider club that uses a winch to launch the sailplanes. By the end of the summer, I took the flight test for the glider instructor rating. I used a borrowed glider and did three tows from a small airport, on the third of which the FAA inspector who was administering the test and I were towed to 4,000 feet. We soared and glided the dozen miles or so to the club glider field where the inspector had me do three winch launches so I would have an unrestricted glider certificate. On the third launch, the inspector stood on the ground and watched while I gave my wife a lesson. A glider certificate is restricted to the kind of launch used on the flight test, aero tow, ground launch, or powered glider.

Throughout that winter I worked over the other two principals in my company, and by spring I had them talked into adding glider training to the programs offered at our flight school. We already had a Citabria which we were using for taildragger checkouts and aerobatic training, and which was equipped with a hook and thus suitable for use as a tow plane, a tug. So for starters we bought a two-place glider to use as a trainer. We later added a single-place glider to our school fleet, and over the next three years we trained quite a number of glider pilots, both from scratch and transitions from power ratings. This glider training program lasted until the airport where we were conducting glider training was sold to a large corporation, which used the property to build a plant, and another nice little airport bit the dust.

Today, of course, there are three ways to get a sailplane aloft: aero tow, ground launch (winch or auto tow), and self-launch, the so-called motor glider. The glider rating on the pilot certificate restricts the pilot to using the type of launch demonstrated on the checkride, and to remove this restriction the pilot must demonstrate to an examiner or inspector his or her ability with each type of launch. If you already have a power rating and are taking a transition course, there is no written examination for the glider add-on.

Keeping the thing up without an engine requires the use of nature's power, and this comes in three basic forms: thermals, ridge lift, and wave soaring. Being a flatland resident of the Midwest, my personal soaring has been limited to using thermals for my lift, except for an occasional expedition to a mountain soaring site where both ridge and wave soaring may be found. One of the greatest thrills I have ever experienced was soaring along a ridge at Dillingham Airfield at Haleiwa on the island of Oahu in Hawaii and looking out over the Pacific Ocean and watching the whales at play.

However, with thermaling alone as my only source of lift, I have made altitude gains on more than one occasion of as much as 8,000 feet, releasing from the tow at 2,000 and thermaling up to 10,000. Unless you've done it, you simply cannot comprehend the thrill of meeting the challenge of climbing in a heavier-than-air machine with nature's power alone. The feeling of exhilaration is literally unbelievable.

One very tangible benefit of flying a glider is this: many of today's pilots who trained in airplanes after the manufacturers started putting a training wheel out in front of the airplane simply have no idea what that rudder is all about. If you want a dramatic demonstration of the principle of adverse yaw, try turning a glider with aileron alone, and watch how the nose swings away from the direction of the turn. Many modern pilots plant their feet firmly on the floor and steer the airplane through the sky like an automobile, with the "steering wheel." And they can get away with this because modern airplanes are designed to permit it, but it makes for somewhat sloppy flying.

It was entirely different when pilots were trained in J3 Cubs and Aeronca 7AC Champs. In those days a great deal of pilot training was concentrated on coordination of hand and foot. Because of the long wingspan of all gliders, the extraordinarily long ailerons, and

ιe down aileron on the up
ces substantially more drag
ι wing, resulting in this ex-
aw. Therefore, the glider pilot
when entering or rolling out of
proved precision in all the pilot's
it. I guarantee that not only will
ill actually enjoy it.

certificate, the transition to gliders is
nent in a transition course is really very
ιng required to learn to coordinate stick
ιy only two new techniques to learn. One,
ɔw the tow plane in a very specific position.
This can be a             at first, and the other new technique for the
power pilot is in lα    ng the glider. One doesn't flare for landing as
is done in an airplane. Rather, the glider is flown right down to the
ground. Once these tasks are mastered, it is simply a matter of taking the checkride, which also includes stalls and emergency procedures, such as a simulated tow-rope break. The glider student must
also learn how to take slack out of the tow rope and to "box the
tow," or fly around the wake produced by the tow plane, and since
there is no written examination for the pilot who has a power rating
once these skills are mastered and the very minimal experience requirement is met, the student is ready for the practical test for the
glider rating add-on. Then comes the fun and challenge of learning
to thermal with the sailplane and gain altitude without an engine.

There is sure to be a glider school or glider club within a reasonable
distance of wherever you are located, so why not go out and take a
ride? Be forewarned, however. If you do, you are likely to be
hooked, just as I was and many thousands before me were.

Personally, I can put my ATP certificate in my pocket, climb into a
pressurized turbine-powered airplane, and get along just fine in the
high-altitude structure where I'm operating as an equipment manager instead of a real pilot, but if I want to have pure fun, I get in a
glider and soar with the hawks. And there must be lots of others
who enjoy soaring as I do, for according to a recent edition of the
"General Aviation Statistical Databook" published by the General
Aviation Manufacturers Association (GAMA), there are many thousands of active glider pilots with glider-only certificates, and countless others who hold certificates with power ratings and glider

ratings. These are the guys who fly for the sheer joy of it. There is absolutely no practical application to flying gliders, but the intangible benefits are immeasurable. If I sound like a nut, it may be because I am absolutely nuts about soaring.

# IV The seaplane rating

Except for the pilots flying in and out of the lakes and rivers of Canada and Alaska, seaplane flying has a very limited practical application. Thus, for the most part it is another example of pure fun flying. And I'm here to tell you it can be lots of fun. To spend a hot summer day splashing around on a series of lakes in a J3 Cub on floats, or a similar airplane, is indeed a joyful activity. I've landed on a quiet lake in such an airplane—wearing nothing but a pair of trunks and a T-shirt—shut the airplane down, and while it drifted gently around, sat on one of the floats with my bare feet dangling in the water and gone fishing. The quiet and peace was absolutely overwhelming.

There are two entirely different kinds of seaplanes, and each must be flown differently. The traditional floatplane is usually a high-wing single-engine airplane fitted with floats, or pontoons. The floats may or may not have retractable wheels, making the airplane an amphibian, capable of landing on either water or land. Those with no wheels are, of course, limited to operating on and off the water and are highly impractical for the average pilot other than for sport flying. These airplanes with floats respond pretty much like they would if they had wheels instead of floats, except that because of the large area of the floats hanging down there, they demand more rudder when entering or recovering from a turn. Also, dragging around all that extra structure makes them slower than their land-only counterparts.

However, the other kind of seaplane, the amphibious hull, is a whole new ball game. Because the engine in a single-engine seaplane of this type is mounted way up on top of a pylon above and behind the cabin, the airplane responds entirely differently from what most pilots expect. When the power is retarded, instead of the nose pitching down, it pitches up. Even more surprising is what happens when you add power. The nose pitches down! This, of course, is exactly opposite of the way a conventional landplane responds to the application of power, and it takes some getting used to. This response to the application of power is frequently quite startling the first time a pilot experiences it. Another factor that is added to the equation in seaplane flying is the addition of the center of buoyancy to the center of gravity and the center of pressure. This factor contributes to the fact that there are three distinct types of taxi operations, idle taxi, plow taxi, and step taxi.

Everything in aircraft design is a compromise. You give a little of this to get a little of that, most notably speed for comfort and economy. And seaplanes are a prime example of compromise in their design. They are neither great airplanes, nor are they great boats, but for the most part they are adequate as either, and whatever they are, they are pure fun to operate. One feels a great sense of freedom as he or she splashes around on the water, and any lake big enough to accommodate you is a suitable landing area. In my own area there are a score of suitable lakes within a 20-mile radius, and I've landed on all of 'em.

Just as is the case with gliders, the transition to add a seaplane rating to the certificate you already have is relatively simple. Again, there's no written examination. It is simply a matter of mastering the skills unique to seaplane operations, such as landing on glassy water (without a good reference by which to flare), rough water, docking, sailing, beaching, etc. Of course, there is a bit of knowledge to acquire when dealing with water operations, such as United States Coast Guard rules of right-of-way, etc. and water etiquette (floatplanes come last, after swimmers, boaters, water skiers, sunken alligators, and everything else in or on the water).

# V Aerobatics

The pilot who learns to fly aerobatics becomes a much better pilot. Just as instrument training tends to make a pilot more precise in his or her heading and altitude holding, aerobatics makes airplane manipulation much more smooth and efficient. Meanwhile, pilots acquire a world of confidence in their ability and that of the machinery they're operating. They gain a whole new insight on the world of aviation. It is here that superstition and fear are truly replaced with knowledge and skill.

It is best to start aerobatic training with small bites, brief sessions, to give your body and system an opportunity to build up to greater and greater G forces. Just as soon as you begin to become the littlest bit queasy, tell your instructor to head for home. After a few sessions you'll be able to take an hour or more of rolling around the sky. When you become proficient at flying precise loops and rolls, you will acquire an extreme sense of accomplishment and satisfaction. Even the most basic aerobatic maneuvers offer a challenge to the pilot and require a good deal of practice if they are to be done properly.

There are a great many advocates of spin and unusual attitude recovery in all pilot training (see Chapter 3). If a pilot has even a small degree of confidence in his or her ability to control the airplane around all its axes, he or she will be a better pilot, certainly a more confident one.

Like glider flying, acro gives the pilot a feeling of freedom like nothing else. If you've never done it, I recommend you get a good aerobatic instructor with an airplane certified for aerobatics and try it. You might even come to like it. Just as instrument training helps the VFR pilot to fly much more precisely whether VFR or IFR, acro training makes a much smoother pilot, an end certainly desired by all of us. And if you should, in your normal flying, blunder into a situation in which you lose control of the airplane and end up in a really unusual attitude, with aerobatic training you'll be able to recover where otherwise you might not.

Like any other skill, aerobatic flying takes constant practice to maintain and improve one's efficiency, and those who enter aerobatic competitions are practicing and improving on a daily basis. I'm not advocating here that everybody should work at it like this, but a bit of aerobatic training does wonders for your ability to manipulate an airplane and consequently for your confidence. Besides, as I said, it's fun! Warning! Do not attempt it on your own. You can't just read a book and go out and do it. Get a good aerobatic instructor and some dual instruction before you try it by yourself.

The IAC (International Aerobatic Club), which sanctions contests, is a division of the EAA (Experimental Aircraft Association), and it, of course, maintains that aerobatics are for everyone. It urges all pilots to add a new dimension to their flying.

## Too old? Portrait of a champion

By law, air-carrier pilots are required to retire at age 60, yet Henry Haigh won the World Aerobatic Championship at age 64! And for anyone who is not aware of it, serious aerobatic competition is one of the most strenuous, physically demanding, and nerve-wracking activities known to humans. It takes a level of concentration unknown to most people, and a degree of dedication only experienced in a very few other human endeavors.

I go to air shows and aerobatic competitions as a spectator, and I watch those guys (and gals) and think to myself, "They are not doing anything I haven't done and can't do. They just do it infinitely better!"

This is in no way meant to take anything away from the air show performers, but competition aerobatics is an entirely different thing. The air show performers fly their routines so well they take your breath away. But they have the whole sky in which to do it. The competition pilots, on the other hand, must stay within the confines of a three-dimensional box or lose penalty points. They are required to fly a series of maneuvers proscribed by the contest committee, a series of their own maneuvers (which are scored by the degree of difficulty), and a surprise routine which they are given just prior to having to fly it.

Henry Haigh is the very smoothest and most precise airplane manipulator I have been privileged to observe in a lifetime of attending air shows and aerobatic competitions that started in the mid-1930s. He got his flight training as an Aviation Cadet in the Army Air Corps in World War II, and after graduation as a military pilot he went on to fly B-24s and B-29s.

In the early 1960s, using a Ryan PT-22, he began to seriously practice aerobatics, and he won the very first contest he entered, and that was at the advanced level! From that time until he retired from active competition in a career that spanned over 20 years, Henry won over 100 aerobatic contests. Throughout this period, Henry was so focused that he put in well over an hour twice a day practicing to achieve and maintain the skill required of a champion. That's the kind of dedication it takes if you want to be a champion. Then there's the unmatched concentration and focus, the nervous tension of competition itself.

After winning the IAC competition at Fond du Lac a total of seven times, placing second in the U.S. championship several times, and winning the United States Aerobatic Championship in 1980, Henry Haigh's career reached a peak in 1988, when at the age of 64 he won the World Aerobatic Championship. (He had placed second in the world in 1980 and 1982.)

Besides being a superb aviator, arguably the best in the world for a 20-year period, what kind of a man is Henry Haigh? Is he a real person, or some kind of cardboard figure for the kids to look up to?

Without a show of false modesty, Henry is self-effacing almost to the point of shyness. He has studiously avoided personal glorification throughout his career, literally turning down a fortune to appear in air shows. To avoid even the appearance of self-promotion, he has steadfastly turned down speaking engagements. After all, his accomplishments speak for themselves. When he does speak, it is with a quiet authority that commands attention. In gathering material for this profile, I had to look elsewhere, for Henry doesn't like to talk of his accomplishments. It would embarrass him to do so.

You will notice that I have omitted from this chapter any reference to rotary-wing (helicopter and gyrocopter), and lighter-than-air (balloons and airships). There are two reasons for this. One, I know absolutely nothing about fling-wing flying, and two, my lighter-than-air experience has been limited to instruction in a hot-air balloon with airborne heater, and quite limited at that. I'm basically a lazy guy, and whereas flying an airplane consists of pointing it where you want to go and letting it take you there, flying a helicopter requires that both hands and both feet be doing something all the time. The damn things are inherently unstable, while airplanes are inherently stable! And as for balloons, they're fun, but they are just too much hard work for me. It's not all that hard to get the thing inflated and flying, but when you land, you have to fold up the envelope, stuff it in the basket, and drag the entire thing out of the field to the road where your chase car is parked, and that's not easy.

Several years ago I undertook balloon instruction and got to the point that I lacked only a single solo flight to qualify for private privileges in a hot-air balloon with airborne heater. Then the guy who was training me sold his balloon, as a result of which I never finished. Even before that time I had bought a couple of hours of dual instruction in a helicopter. Talk about balancing on a beach ball! The guy had me hovering a few feet off the ground, and I had both hands and both feet going full time, while some idiot (control tower operator) was chattering away at me on the radio expecting an answer! No thanks! I decided this wasn't for me. Too much work. Besides, one of the instructors who worked for me (teaching primary students in airplanes), and who had 800 hours of experience flying helicopters in combat in the jungles of Vietnam had this to say about fling-wing flying: "If you encounter a problem in an airplane, you just glide down and land on some suitable surface. If you have a problem in a helicopter, it flings itself to the ground and thrashes itself to death!"

# VI  A helicopter story
## Prelude

Since I don't know enough about helicopter flying to give you any really useful information, I'll herewith pass on what I believe to be an interesting helicopter story. Jerry Temple is that rare individual, an exceptionally honorable airplane broker, specializing in brokering twin Cessnas. He is based in Carrollton, Texas, and is absolutely the world's expert on the entire line of twin Cessnas. Anyone dealing with Jerry knows exactly what he's getting and what he's paying. In discussing flight testing, which is my area of expertise, Jerry told me of a unique experience he had on a checkride a few years ago.

## The event

The story starts in Vietnam in January of 1967. At that time Jerry was a 21-year-old army helicopter pilot, flying a Bell UH-1D (Huey) Utility Helicopter for the 227th Assault Helicopter Battalion of the U.S. Army Air Cavalry Division. On the day in question, Jerry had been flying lead ship in a flight of two Hueys throughout the morning hauling supplies (food, water, munitions, mail, other assorted equipment, and an occasional passenger) to troops in the field from their base camp, Landing Zone English.

At about 1:00 p.m., after having worked at cargo hauling all morning, the flight of two was returning to LZ English for lunch. When about 15 minutes out from English, they heard a distress call on the radio. "Mayday, Mayday, Chinook hit. We're heading for a sandbar in the Bong Song River about 25 miles Northwest of English!"

The Boeing VERTOL CH-47 Chinook is a twin jet (tandem rotor aircraft) and was the First Cavalry's big cargo and heavy-lift helicopter. They were considered very valuable to the Cavalry and a high priority existed for CH-47 defense and field rescue and recovery. Jerry Temple's flight of two was at that time heading home about at 1,500 feet and doing 90 knots. Since they were the closest U.S. troops, they headed for the downed Chinook.

While Jerry flew the lead Huey, the Aircraft Commander of the other one was talking to Landing Zone English, calling for helicopter gunships, tactical air support (Air Force, Navy, and Marine Corps jets), and, most important, troops. After all, Jerry's flight was just two unarmed Hueys that fortune had placed nearest to the wounded Chinook. Meanwhile, they were learning from the downed Chinook that only crew was involved. This, says Jerry, was both good and bad. They had no troops to set up a defense, but since the crew numbered only five, they could be evacuated on the two Hueys.

Following the numerous bends in the river, Jerry descended to 50 feet and 95 knots with his eyes outside the cockpit while the non-flying pilot worked the radios and checked the gages. Coming around a final bend in the river, they spotted the big green Chinook sitting on a sandbar located fairly close to one bank. There was, however, room for the two Hueys. In Jerry's own words, "A running landing on hard ground was routine, but wet sand that you've never set foot on can cause problems. As the pilot flares the helicopter, he puts the aft end of the skids into the sand and gently rocks forward on the skids so the aircraft is level and is slowly sliding forward. It really happens very fast. The risk in this technique, according to Jerry, is if the sand is very wet and slushy and the pilot sets down at too great a speed, the rear end of the skids are prone to sink in, and at a fast forward speed, as the pilot attempts to level off, the helicopter is likely to wheelbarrow up on the forward skid toes, causing the rotor blades to contact the ground right in front of the helicopter, a most undesirable result indeed!

Temple briefed his crew, calling for, "Belts and harness tight and helmet visors down," standard Cav call-outs. He advised his wingman on the company frequency of his intentions and told him to keep the formation in tight. He also advised his gunners to load and go unsafe. About this time, some 50 yards from touchdown they observed light flashes from the shoreline and the Chinook reported that they were taking fairly heavy light-arms fire. Little eruptions of sand were popping up close to the CH-47. All the personnel in the downed Chinook and the two Hueys were hoping that the fire would continue to be small arms and not become rockets or mortars. Just as Jerry completed the quick-stop maneuver, his right-door gunner opened fire and shouted, "Taking fire from 2:00 o'clock!"

Obviously the enemy had held off until help arrived for the downed Chinook before letting go with all they had. Jerry says, "They had gambled that we'd not had two loads of troops on board. They (already knew) the Chinook crew was solo....Though one CH-47 and its crew was a prize that would get you medals, a Chinook and a Huey or two would make you a true 'people's hero'." The big Chinook had set down just across from a village in a hostile area, and a CH-47, along with two UH-1Ds and 13 Americans, were quite a gift.

The bullets hitting Jerry's Huey were going "ping, ping," nothing like the sound on TV and in the movies. I guess the ping and crack sound just isn't dramatic enough for the producers of the films. With this small-arms fire from the shore kicking up sand all around them, the five-man crew of the Chinook came spilling out of the rear cargo door, carrying the two large machine guns and ammunition that was their only armament.

It was still only small-arms fire from the shore, but as Jerry points out, your brain doesn't categorize the size of the munitions being fired at you. These people really mean to kill you!

Two of the Chinook crew members scrambled aboard Jerry's Huey with a machine gun and their personal weapons and other gear while the other three boarded the second Huey, and the order was given to "Go!" and both Hueys took off and flew directly over the village from which the fire was coming. By this time the enemy realized that they should have nailed the Chinook and crew immediately instead of waiting to ambush the rescue party. Both Hueys had been hit several times, but both were flying okay as they flew clear of the area, and no personnel were injured.

Jerry's flight of two Hueys was safely on the ground at Landing Zone English within 20 minutes. It was time for lunch prior to continuing the day's trucking with replacement aircraft. Meanwhile a strike team would be annihilating the village while a recovery team would be bringing out the Chinook.

## The checkride

After Vietnam, Jerry Temple served as an Army instrument instructor pilot and examiner authorized to issue military instrument ratings in both helicopters and T-42s (the military version of the BE-55 Baron). However, 20 years after "the event" Jerry had been out of flying for four years and he missed it. Consequently, after an intensive period of review, he landed a part-time FAR Part 135 job flying a Cessna 414A. This, of course, required a 135 checkride, and Jerry hoped to combine this with the reinstatement of his CFI certificate with all ratings.

This posed a problem because he had to locate an inspector or examiner authorized to administer both the 135 check and to reissue the instructor certificate with all the ratings that Jerry held (A-S & MEL, Instrument Airplane, Rotorcraft-Helicopter, and Instrument Helicopter). However, after much calling around, begging and pleading, Jerry finally found an examiner qualified to do it all and an appointment was made. The examiner was not pleased with the idea of combining it all into a single flight test, but he agreed to consider doing so.

The cost of renting a twin for the CFI reinstatement was daunting, as was the thought of additional preparation, so Jerry hoped to get it all done with one transaction. He advised the examiner that they would do the Part 135 ride and then address the instructor situation as appropriate.

When he advised the Part 135 operator and the flight school—where he had accomplished his preparation in a simulator and a CE 310—that he had a appointment for the ride, they were pleased. But when he identified the examiner with whom he had the appointment, they reacted in his words, "as though he had crashed and burned." He was informed that no one had ever passed a Part 135 ride with that fellow the first time out. He was retired military and an extremely tough examiner who didn't like civilian aviation and civilian pilots. He was told to cancel the appointment and look for another examiner.

However, his ego wouldn't permit him to do this, and feeling that he was well prepared and ready, he kept the appointment not really knowing what to expect, but he had given enough checkrides and taken enough checkrides to feel confident.

He arrived early at the small FBO where he was to meet the examiner, positioned the 414A on the ramp, and laid out all his material on a table. When the examiner arrived, they discussed Part 135, IFR procedures, ME philosophy, 414A performance, etc. It was then time to go out for the flight portion of the test. Jerry asked about the instructor reinstatements, and the examiner replied that the 135 ride would take up the entire afternoon, which Jerry interpreted as a negative on the instructor stuff.

As they were doing the walkaround preflight of the 414A, a big green National Guard CH-47 Chinook passed overhead, and Jerry mentioned that they were a great heavy-lift helicopter.

The examiner asked him how he knew about Chinooks, and Jerry informed him that he had been a Vietnam-era Army helicopter pilot, flying Hueys in Southeast Asia. The examiner asked him when he had been in Vietnam and Jerry responded, "In late 1966 and 1967. I was with the First Cav. Who were you with?"

The examiner said he was also with the First Cav at that time in the 228th, the First Air Cavalry's CH-47 Chinook Battalion. Jerry then said, "Well, if you flew Chinooks for the 228th in early '67 you probably remember the Chinook that went down on a sandbar in the Bong Son about 25 miles from English?"

"Remember?," he said, "Hell, I was the Aircraft Commander!"

Jerry reports that on hearing this, a pleasant feeling surged through his entire body as he explained that he was the guy flying lead ship in the flight of two Hueys that arrived within a few seconds of the Mayday call from the Chinook. "You're looking at Rattlesnake 39. Nice to see you again. Let's go fly."

The checkride was thorough and fair. Jerry was ready. He passed. He was Part 135 PIC approved. His instructor certificate with all ratings was reissued. Which all goes to prove that a well-prepared applicant should have no trouble passing a flight test, but you never know what else might help a bit.

# 11

# The air traffic control system

If you fly IFR it is extremely important that you be comfortable in the system. Although all airplanes respond to the same immutable laws of nature, when the weather is really grim and you can't even see your own wingtips, there's nobody up there but you and the pros, the people who earn their living by shoving tons of metal around the sky. And to mix a metaphor, it's a different swimming pool, and if you want to play in their ballpark, you'd better know what you're doing because they surely do! I have had many instrument students tell me that flying the airplane on the gages is easy, but the hard part about acquiring the coveted instrument rating is learning to live in the system. I'm sure you've been told many times that the last thing you need to do is talk to the people on the ground. They're not going anywhere. If you get in a big hurry to answer them and you get in trouble because you were talking instead of flying the airplane, the result could be disastrous to you, and they'll just have another cup of coffee while watching the next airplane on the scope. Your primary duty is to fly the airplane. The thing to remember about the people in ATC is that they are just that—people, and they respond just like other people do.

The center or approach controllers' job is to move you along while keeping you away from other known traffic. They want to get you out of their hair as soon as they possibly can so they can deal with the next pilot. Therefore, they're going to pass you off to the next controller down the line just as quickly as they can.

## ∎ Flight service

Flight service stations are a sort of stepchild of the Air Traffic Control System. In the collective mind of the public (and even many aviators), ATC means ATCT (Air Traffic Control Towers) and ARTCC (Air

Route Traffic Control Centers), but flight service stations are really a part of the ATC system, and a very important part, for it is with the flight service specialist that the average pilots have their first (and possibly their only) contact with ATC. It is here that the basic VFR pilots get to talk with a real, live, breathing human being who can explain the weather to the pilots and answer their questions.

Since the consolidation of the flight service stations it is not always possible to do so, but student pilots should try to visit an FSS with their instructors. This is frequently the only contact the average VFR-type pilot gets to have with the ATC system, and it is quite a valuable one indeed. Personally, I think it was a great mistake for the FAA to close down so many flight service stations and put the money into such things as parking lots at major air-carrier airports when a personal visit to the local FSS did so much for the pilots. DUATS is wonderful, and the telephone is a great gadget, but nothing can replace the personal contact of a visit wherein the pilot has the opportunity to talk with a real, live, breathing human being.

If you want to get aviation weather, don't bother talking with the Weather Bureau. Those folks are used to answering questions such as, "Will it rain tomorrow afternoon when Aunt Suzy is planning a picnic?" The Flight Service Specialist, on the other hand, is trained to deal with pilots and to interpret weather charts, sequence reports, prog charts, and forecasts for them. And whether it be over the telephone or a personal visit, this is usually the first contact a pilot has with the FAA.

I personally believe that on the whole, with very few rare exceptions, ATC personnel are the greatest public servants in the history of the human race. These guys (and gals) will almost invariably knock themselves out to serve the flying public. And the best of the bunch are the flight service specialists. The patience and understanding of these people when dealing with the questions of a bunch of poor dumb pilots is nothing less than marvelous. If you've never visited a FSS (and many instructors don't bother to take their students to one when on a dual-student cross-country trip), by all means do so. And if the last time you went to one was a few years ago and you've never been to one of the fairly new automated jobbies, go and get the tour through one of them. It will really open your eyes and mind as you see some of the great new technology these good people have to work with.

One excellent service formerly provided by the FSS but no longer available is the DF Steer. By this technique a VFR pilot caught in IMC was able to get assistance not only to an airport, but through a complete approach to a safe landing. At many locations it is still possible to get direction-finding help from flight service, but the approach feature is no longer available. If you are located where this service is available, by all means ask for a practice DF Steer, comply with the instructions of the FSS specialist and land at the airport where the flight service station is located. Then go in and ask to speak with the people who were working you in the air. They'll show you exactly how they plotted your course on an acetate overlay on a chart on the desk and the scope they used to find and direct you. I guarantee you will find it an extremely interesting experience. I still do this with all my primary students, and all of them are delighted with what they learn.

Some flight service stations have an EFAS (flight watch) position. This operates on the common frequency of 122.0 MHz and exists to give airborne pilots real-time (current) weather information. It is not to be used for any other purpose, but it is a wonderful service to have available to us. By simply monitoring the frequency, you can get all kinds of valuable information, and as pointed out in Chapter 9, it is always a good idea to get current weather at your destination anytime you are on a 30-minute or longer trip.

## II The DF steer

How many of you reading this have experienced a DF steer? Let's see a show of hands. Okay, you can skip this section, both of you. The rest of you, pay attention! Prior to the consolidation of the flight service stations, it was much more common, and even included approaches designed to get a lost or disoriented pilot, or a VFR pilot who had blundered into IMC, safely on the ground. Now it is a technique used primarily to help lost pilots locate themselves. The DF stands for "direction finding."

It works like this: when pilots confess their situation to a flight service specialist, usually on 122.2, .3, .6, or 122.1 while monitoring the nearest VOR frequency, the FSS specialist will first give the pilots a discrete frequency so they won't be bothered by others using the common frequency. The specialist will then ask a series of questions, some of which are designed to put the pilots at ease. If the airplane is equipped with a VOR receiver, the pilots will be asked to tune a specific VOR, center the needle on the OBS with a "from" indication, and tell the FSS person what it says. The pilots will then be told to key their mic and hold it for a specific number of seconds.

This will shoot a line on an oscilloscope at the flight service station from the airplane to the FSS antenna. By transposing this line onto a chart, and drawing a line from the selected VOR, the FSS specialist has now, by means of triangulation, located the airplane.

The specialist will then give the pilots a heading to a nearby suitable airport. After a few minutes, giving the airplane time to travel enough distance to determine the actual track over the ground, the FSS specialist will repeat the triangulation process and thus learn how far and in what direction the airplane has traveled from the last known location since the last triangulation.

With GPS, Moving Maps, Loran C, and all the other wonderful stuff we have today, the DF steer is gradually falling into disuse. However, for the student pilot in a primary training airplane with only basic equipment, it is still a valuable tool, and one ignored by most flight instructors. In fact, many flight instructors have never even heard of it.

## III  The tower

Those airports that have enough traffic to warrant a control tower may have either a full- or part-time tower. A full-time tower is in operation 24 hours a day, every day of the year, while a part-time tower offers limited hours of operation. In either case there are at least three positions which are filled, but in many cases, two or all three are combined and one individual covers them all.

The three basic positions in the tower are: one, local control. This individual is responsible for all traffic in the air in the immediate vicinity of the airport, and anything on the active runway. The local controllers are the people who clear you to take off and land. They're also the people who actually put you into the IFR system when they hand you off to either an approach (departure) facility or an Air Route Traffic Control Center. The second of these positions is ground control, and these people are responsible for all movement on taxiways, and some ramp areas, up to but not including an active runway. Remember, a clearance to a runway authorizes you to cross all intervening runways (unless told to "Hold short of....") but not onto the active runway. Finally, there is a position in the tower called flight data. The responsibility of the individual working the flight data position is to coordinate IFR departures and arrivals with the next controller down the line, which is usually an approach control facility. The person on flight data also passes tower en route, or tower-to-tower flight plans along. All this is done on the land line (telephone). The individual working flight data also handles clearances to and from the first en route controller, either an approach (departure) or a center controller.

There is, in some very busy towers, a clearance delivery position. In less busy towers, this function is handled by the person working flight data, and in those towers with even less traffic by the ground controller. All primary airports in Class B and many in Class C airspace offer pretaxi clearances, which serve to get the pilot on his way after takeoff. These clearances are issued whether one is IFR or VFR, and it is important to remember to switch over to ground control after receiving a clearance prior to taxiing out in those instances when a pretaxi clearance has been issued, a procedure with which VFR outbound pilots occasionally fail to comply.

# IV Approach and departure control

Occupying a position between the local (tower) controller and the center controller is the RAPCON (radar approach controller). When the tower hands off a flight and requests the pilot to switch over to "departure," this is the facility to which the pilot next talks. In other words, approach and departure control are the same person.

Every square inch of the continental United States belongs in the jurisdiction of one or another Air Route Traffic Control Centers, or Centers as they are called. Boy! do I hate that term. They use the expression "owning" this or that piece of the airspace. They don't own it; we, the citizens, own it! From this huge amount of sky are carved out columns of air, which by letters of agreement are assigned to one or another approach control facility and/or Air Traffic Control Towers.

The column of air around a tower goes from the surface up to 3,000 feet above the surface. These columns assigned to approach/departure facilities extend up to 6,000, 8,000, 10,000, or 12,000 feet above the surface and are found around busy airports serving metropolitan areas. They are classified as "B" and "C" airspace. The RAPCON (radar approach control) or TRACON (terminal radar approach control), while having absolute control over the column of air assigned to it, has radar coverage that extends far beyond the borders of the airspace it "owns." This permits IFR hand-offs from Center (or the adjacent approach facility) far enough out for the traffic to get into position for an instrument approach.

On those occasions when your entire trip is in airspace covered by approach facilities by virtue of two or more adjacent facilities whose geographic boundaries physically contact each other, you may file an abbreviated flight plan with the tower at your departure airport and get what is called a tower-to-tower, or tower-en-route clearance. In those cases, the pilot only talks with approach control people, never directly contacting the center. Each approach facility has direct land-line contact with all adjacent facilities and the center.

These are usually the first people a pilot talks with after being handed off by the ATCT (air traffic control tower) on departure and the last before being turned over to the tower on arrival. Of course, if you're departing from or arriving at a nontower airport out in the boonies, you may never get to contact an approach facility, since your entire trip may be in airspace "owned" by Center. As is the case with all ATC facilities, their function and only real responsibility is to separate known IFR traffic, but they do a great deal more. If requested politely, no matter how busy they are, they will usually provide advisories (flight following) to VFR traffic in their area. Flight following is another great service offered to us pilots, and I take advantage of it every chance I get. It does not relieve the pilot of responsibility to see and avoid traffic, but it does provide another pair of eyes watching out for you. I don't know about you, but personally I want all the eyeballs I can get helping me look out for traffic. In such a radar environment, IFR flights are invariably vectored to the final approach course, saving pilots from having to maneuver themselves around to get lined up for the approach by flying the entire procedure—procedure turn, DME Arc, or whatever.

# V En route

All the rest of the controlled airspace is "owned" by one center or another. For a good many years, ever since the phase out of the old colored airways, we have been operating, in the low-altitude system (below 18,000 feet MSL) on victor airways, primarily defined by radials from VORs and VORTACs. These airways, like our highway system, are given odd numbers when they run north and south, and even numbers when their direction is east-west. With the advent of the Course Line Computer (RNAV), Loran C, and GPS (Global Positioning System), we now have the capability of going direct from departure to destination. Air Traffic Control, however, has not been very quick to accept this new direct method of navigation. Our airplanes were equipped for direct navigation before ATC was ready to handle it. For several years we were kept on airways even though we had the capability of going direct.

In a nonradar environment, an IFR approach clearance will be issued by the center, and in that case the pilot must fly the entire procedure, which may include a procedure turn, a holding pattern, or a DME arc to get lined up with the final approach course, and in some situations a straight-in approach from a nearby facility (a VOR or a nondirectional beacon). And the DME arc is another of the things that is often not taught anymore. If you are an instrument-rated pilot and are not familiar with this procedure, get yourself an instructor, find an airport that has an approach using the arc as a means of joining the final approach course, and learn how to do it. Where appropriate, it is an extremely useful device, and it is important for every instrument pilot to know and understand this procedure. While not particularly difficult to execute, it does take some thought and practice to perfect.

Although ATC personnel don't particularly like to accept en-route air filing of IFR plans, or "pop ups" as they are called, traffic and workload permitting, many approach and center controllers are happy to issue them. The key is to ask politely. You know the routine. "We're unable to maintain VFR. We're rated (instrument rated and current) and equipped (the airplane is properly equipped for IFR operations). Traffic and your workload permitting, I wonder if you would be kind enough to clear me, present position to... (destination)." Unless the controllers are very busy, or in bad moods, they will almost invariably comply with your request. Using that kind of language for your requests, you can usually get just about anything you want. On the other hand, if you come on with unreasonable demands and insist that a busy controller do something or allow you to do something out of the ordinary but which is perhaps more convenient for you, you are likely to get an unwanted tour of the entire area.

# VI Clearances

I'm a sort of lazy guy. I don't like to do any more than is required. Consequently, I always like to hear the controller say, "Cleared as filed, contact Departure (or Center) on frequency... Squawk..." or "Cleared to...via flight plan route. Maintain...expect filed altitude ten minutes after departure...Contact Departure (or Center) on frequency...Squawk..." Did you notice there was nothing in there about a route? That's because I make it a point to look up the preferred route and file for that. Since that's all the computer at the center knows, it is what I'll get anyway, so I might as well file for it even if there is a much more convenient route to my destination. It is impossible to reason or negotiate with a computer. But once in the air you are dealing with a human being, and if you ask nicely, you will no doubt get what you want. Meanwhile, I have avoided writing down a long, involved route simply by anticipating what the computer will spit out.

I fly fairly often from Pontiac, Michigan (PTK) to Cuyahoga County, Ohio (CGF), and for many years the preferred route would have taken me over an intersection smack dab in the middle of Lake Erie. Added to this is the fact that the RAPCON has a policy of keeping the "little airplanes" low (at 3,000 feet) until they are well past the departure and arrival routes for Detroit Metropolitan Airport (DTW). Now, in addition to being lazy, I'm chicken. I am willing to cross Lake Erie in a light single-engine airplane if two conditions are present. One, I want to be at 9,000 feet with everything stabilized before I start across, for it is my belief that if anything is going to happen to the engine, the odds are overwhelming that it will happen when a power adjustment is made. The second condition I require for crossing the big water is that I fly south by the islands. There is a string of islands near the western end of Lake Erie, and there is an airport on every one of them. (One even has a published instrument approach.) Going that way, one is always within gliding distance of land if he or she is at least 9,000 feet up.

Knowing all this, I would file for the preferred route, which would take me right out in the middle of the lake, and I would accept the clearance on the ground. However, as soon as I got into the air and in communication with the RAPCON, who wanted to keep me at three, I would start screaming for higher, but I would explain that I wanted to be at nine before starting across the lake, and I would invariably get it. Then, as soon as I was handed off to Center, I would ask for "direct Sandusky" (the SKY VOR). This would take me across by the islands, and I would again explain why, and I would again get just what I was requesting. I wouldn't go to Sandusky because as soon as I was within gliding distance of the south shore of the lake, Cleveland approach would be vectoring me toward Cuyahoga County Airport (my destination). The moral of this story is: although you can't argue with a computer, you can always negotiate with a human being, and you'll get what you want if you make reasonable requests and explain yourself.

It is well known to one and all that the use of a cellular telephone in an airplane is a no-no, but that's not the entire story. It is against Federal Communication Commission regulations to use a cell phone in an airplane in flight, but not on the ground. Therefore, if you are at a nontower airport and are fortunate enough to have a cell phone, you can avoid the pressure one always feels when he or she has a void time window. With your cell phone, you can sit at the end of the runway and talk directly with the RAPCON or center and get your departure release without the hassle of having to squeeze out in the window of time offered by a void time clearance. Neat, huh?

# VII Operation rain check

I'm sure you are familiar with the aviation safety seminars, also called accident prevention programs, or pilot education programs conducted by the local Flight Standards District Office Safety Program Manager (SPM), formerly Accident Prevention Program Manager (APPM), and before that Accident Prevention Specialist (APS), or one of the civilian volunteer counselors. As an aside, I should tell you that the FAA is in love with words. They insist that we all use the recommended words, but then they keep changing the words. I believe they quit calling those people apes in order to make them feel more important. After all, they next had "manager" in their title. Then, since the word "accident" in accident prevention has a bad connotation, some smart official decided it would be more positive if they were called Safety Program Managers, since "safety" is positive and "accident" is negative. Thus, they are now SPAMs. In any event, each of the programs put on by these fine men or women, or one of their counselors, focuses on current safety-related topics, and they usually address specific concerns.

These programs, or Pilot Education Seminars as they are sometimes called, are now part of the "Wings" program. If you are not familiar with and involved in this terrific safety program, by all means hie yourself down to your local FSDO and get all the details. (Or call 'em up and ask for the AC on the Wings program.) Then become involved. Being the safety program manager at a FSDO has to be one of the most frustrating jobs in the world. People like to be able to measure their accomplishments, but how can people possibly know they have prevented an accident? Well, for the first time in my knowledge, there was a means of knowing that a positive result came out of an accident prevention program. After the first full year of the operation of the Wings Program, not a single one of the 20,000-odd pilots who had earned their wings had been charged with a violation or been involved in a reportable accident or incident! That, to me, is an impressive record. Of course, the pilot who is interested enough to be involved in such a program is the kind of person who is not likely to commit a violation or be involved in an accident or incident anyway, but it is still an impressive record. We once had an APS who knew how to measure the results of his effort at preventing accidents. I was in his office at the FSDO one day, and every once in a while he would reach up and put a hash mark on his wall with a piece of chalk. When I asked him what he was doing, he replied, "I just saw an airplane land out there on the runway, so I know I have prevented another accident!"

Unlike any of these Flight Standards safety programs, Operation Rain Check is a safety education initiative of the FAA Air Traffic Service, which focuses on airspace and air traffic procedures training for those of us who use the system. These programs are conducted by air traffic controllers from local towers, approach control facilities, and even Air Route Traffic Control Centers.

It is at these Operation Rain Check seminars that pilots have the opportunity to learn about the unique procedures used by air traffic controllers in the geographic area in which they operate. It is an excellent opportunity to find out why certain clearances or procedures are used within the specific area, and these programs offer pilots a chance to voice their concerns regarding specific air traffic problems, and, best of all, to place a face with a familiar voice on the radio. Getting personally acquainted with the controllers in your area is very important. After all, the largest segment by far in the FAA is the Air Traffic Service. I have found the controllers to be extremely cooperative and willing to change a procedure if you can show them a valid reason for doing so.

It is this segment of the FAA with which pilots most frequently deal, but other than through formal radio contact, pilots and controllers rarely encounter one another. By attending an Operation Rain Check in his or her area, a pilot learns about radar arrival and departure sequencing, tower hand-offs, and clearances issued for particular operations. This program provides a great opportunity for pilots to learn ways to help themselves and their fellow aviators and for ATC to use the system to best advantage.

When controllers bring diagrams and charts of their areas of coverage to the Rain Check program, they describe just how they cooperate with neighboring sectors and airspace "owned" by the center. This enables the pilots to understand just how they fit into the traffic in the big picture. The best Operation Rain Check programs are those put on by the centers themselves. Once the pilots thoroughly understand the system in which they operate, they will know just what to ask for to avoid unnecessary delays and frustrations. In addition to all this, the pilots learn how such things as special-use airspace and military training routes are used. I am without a doubt one of the most polite guys using the system. I always make reasonable requests, never unreasonable demands, and invariably I start off, "Your workload and traffic permitting, I wonder if you would be kind enough to check with the next controller down the line and see if I can get direct to so and so?" Works like a charm. I almost always get whatever it is I'm asking for. Try it, I guarantee success. And when controllers goof, I apologize. They know they're wrong, and you can't believe what wonderful treatment I get next time I'm talking with these same controllers!

To get a Rain Check program set up in his or her area, all a pilot has to do is contact the Air Traffic Manager, be it a tower, approach facility, or center, and make the request. These people are happy to make this program available because it is not only good advertising for ATC, but also it furthers cooperation between the providers and the users of the system.

# VIII  How to get what you want

None of us really know how we sound on the radio, but apparently I have a very distinctive voice, for controllers all over the country know and recognize my voice. The control tower at the airport where I am based is a training facility for newly hired controllers, and when they are well qualified, they move on to larger facilities where they can earn greater pay. Thus, wherever I go I am likely to encounter a controller who knows me, at least by voice.

As mentioned previously, I am extremely polite and cooperative. By starting out with, "Traffic and your workload permitting, I wonder if you would be kind enough to see if I can have..." I invariably get whatever I'm asking for. Of course, it helps to know what you want. In other words, keeping in mind where you are and what you want to do, carefully think through the best means of accomplishing your objective, and then ask politely. With very rare exceptions you will get just what you request no matter how busy the controller is. In other words, the more specific your request, the more likely you are to get what you're asking for. If, for example, you make a general request such as, "I'd like a more direct routing," you are likely to be kept on the undesirable routing you're already on, but if you say, "How about a heading of such and such to join the airway after such and such VOR?," traffic permitting, you'll probably get it. Bear in mind the fact that the controllers want to get rid of you as soon as they can. Therefore, they will do whatever they can to expedite your trip across their scope so they can pass the problem (you) on to the next person down the line.

It seems that the attitude of all the controllers at a given facility filters down from the top. The facility manager sets the tone. Also, the response you get from a controller is somewhat determined by the amount of traffic that particular facility handles. For example, Cleveland and Chicago Centers are quite busy, and so their controllers are very businesslike, while Minneapolis and Albuquerque Centers are much more casual. (So help me, it seems like if there are two airplanes in the sky controlled by Minneapolis Center, as much as 600 miles apart, they think they have a traffic problem.) What I am basically saying is that if you want cooperation from the other person, you should be cooperative yourself. In other words, make requests rather than demands.

That you can't always get what you want, or even need, is illustrated by the following story. As I said, the uncooperative controller is the rare exception, but a few years ago the Atlanta Center airspace was an absolutely terrible place in which to operate. It didn't matter if there was no traffic at all, you were going to be made to fit into their system where and how they wanted you, no matter what. It is my usual policy to fly from Michigan to Florida (especially the Tampa Bay area) by the Western route with a fuel stop at Knoxville, Tennessee, rather than the eastern route. The highest MEA (minimum en route altitude) is on a segment that goes from TYS (Knoxville) directly over Atlanta at something like 10,000 or 11,000 feet. I have never gone this way without being vectored all around the outer limits of the Atlanta TCA (now Class B airspace).

One clear night (bright moonlight and all the stars out) at 2:15 a.m. as I was returning to Michigan from Florida via this route, when I entered Atlanta Center's airspace, I started getting vectored. I was at 8,000 and there is an intersection just north of Atlanta with a crossing altitude of 9,000. I began to beg to be allowed to climb to niner as I was being sent 50 (that's right, 50) miles east. The controller held me at 8. Believing there must be a reason for sending me on this long expedition and holding me down at 8,000, I scanned the sky looking for the traffic that might have justified the controller's action. Remember, the night was crystal clear. There wasn't another airplane in the sky. I ran through all the other frequencies the controller might be using to see if the controller was talking to anyone else, and I found nothing but silence. Finally, after having been waltzed all over the sky, when I was almost on top of the intersection that required me to be at 9,000, the controller let me climb, then handed me off to the next controller down the line.

Instead of my usual, "(Repeat the new frequency), thank you and good day to you, sir." In a very nasty tone I sarcastically said, "Before I leave you, I want to thank you for the unnecessary 60-mile tour of the Georgia countryside!" My wife, who had been semidozing in the middle row of seats behind me, hearing all this on the speaker, said, "You can't talk to him like that!"

My reply was, "Oh yeah? I just did!"

About then we got handed off to Knoxville Approach Control, and when I checked in with, "Good evening, Knoxville Approach. Here comes 8744 Echo, with you level at nine." The controller came right back, "Forty-four Echo RADAR contact. I see you're the guy who told 'em off at Atlanta! Good for you!"

From there on, all the way home, every time I checked in with a new facility (Lexington, Indianapolis Center, Cincinnati Approach, and Cleveland Center), every single controller congratulated me for telling 'em off in Atlanta!

# IX  Responsibility (again)

Wonderful as the Air Traffic Control system is, it is still the pilot's responsibility to fly the airplane. Many pilots, both veterans and recently rated instrument pilots, simply put themselves in the hands of ATC and expect the controller to keep them out of trouble. In other words, they quit flying the airplane and merely sit there fat, dumb, and happy while the controller tells them what to do. They have become passengers in their own airplanes.

And this is an easy trap to fall into. For many years, until he got promoted and we no longer hear his voice on the mic, we had one of the greatest controllers in the business here in our area. Paul has a distinctive voice that absolutely reeks of confidence, and no matter how bad the situation was, every time I was handed off and heard Paul's deep voice, a great sense of security came over me. I just knew everything was going to be all right. After all, Paul's in charge now! Of course, he really wasn't. I was still responsible for flying the airplane and getting it safely back on the ground, but he made it easy to do.

The point we must all keep in mind is that the controller can't crawl into that cockpit and fly the airplane for us. As pilots we simply must rely on ourselves to make the hard decisions and to act appropriately.

It is imperative that when we call "ready" (for takeoff), that we be ready—somebody might be coming. And even though a controller has cleared us for takeoff, it is still our own responsibility to look around and ascertain that there is not another airplane on final, about to land on our runway of departure. This does not mean that I recommend that you sit there and look around before moving, but, rather, that you look around as you taxi slowly into position on the runway (see Commandment 3 below).

# X  The 10 commandments

NATCA (The National Air Traffic Controllers Association) has published 10 commandments for pilots, setting forth the rules for dealing with controllers.

1. Turn from thy appointed way hurriedly when instructed to by the controller, lest thee find thyself making merry with thy fellow birdman's appendage...for the controller's sight encompasses that which thine eyes cannot see.

2. When the controller sayeth unto thee, with the voice of urgency "HOLD," holdest thou, with the GREATEST of urgency and WITHOUT argument...lest this be thy FINAL opportunity for thee to hold.

3. Should the voice from the air, which is the controller's, clear thee for takeoff, GO LIKETH THE WIND...for perchance there is a machine of flight on short, short final which planneth to use the very surface upon which thou sitteth in a very short time...yea, even unto seconds.

4. Should conditions surrounding thee be that which are known as IFR, asketh him not for a VFR takeoff..should he allow for it, he will find himself in sore trouble with that agency known to him as the FAA, and the law of the land adjureth harsh penalties upon these happenings.

5. Speaketh unto him, thine controller, with a voice of honey, useth him as a brother, lest he become excited, confused, loseth his wits, and giveth thee a right turnout when a left turnout benefiteth the occasion...for, lo, a controller loveth a calm, courteous pilot above ALL things.

6. While in his area, keepeth thine controller informed well in advance of thy every move and intention, and believeth not that he readeth thy mind, for in spite of popular opinion, he is, alas, human even as thee and me...

7. When thou hearest the words from the black box saying unto thee, "UNABLE TO APPROVE DUE TO TRAFFIC," beseecheth thee not from thy lofty position to challenge his decision, for lo, had not the traffic been there...the words would not have been uttered, for he has the eyes of the eagle and sees all WITHOUT fail.

8. When the clearance is of the VFR type, stay ye from the proximity of thy brothers holding, for lo, the poor controller is sorely tried to explain to his IFR charges the strange birds in their presence.

9. Asketh for instructions in a voice that is calm and clear, so thine controller will understand thy wants and desires; confuseth him not lest he clear thee for landing on runway 13 while clearing one of thy brothers for takeoff on runway 31...

10. Watcheth thou closely for all four-wheeled vehicles. They are numerous and unpredictable; yea, even as the whirlwind. Treateth them with respect and fear while taxiing lest they

charge upon thee with the speed of the lion and the fury of
the tornado...for their drivers may be uninstructed in the
ways of the birdman. So pity the poor controller for his
troubles are many, and the transgressions against him are
many...even into the thousands, yea verily. Therefore, show
thy mercy unto him, for he is sorely tried. He acteth as a
guardian angel to the poor misguided birdman, and in return
receiveth harsh words, unkind looks, and has all manner of
evil happenings bestowed upon him. His very acts are
guided by the handbook known as 7110.65; and should he
transgress therefrom even to lo, one misapplied portion of
the phraseology, all hands revile him and make light of the
times, yea even to the millions of times he has been right.
Therefore, I say unto thee, honor thy poor controller and
heed well these 10 commandments, that thy days in the sky
be of long duration...yea verily!

# 12

# Checklists, other paperwork, and certificates (privileges and limitations)

## I Passenger briefing, seat belts, and harnesses

Back in Chapter 1 we discussed prop safety and how important it is for pilots to warn their passengers regarding the danger of walking into a spinning propeller. Not only is this necessary with regard to disembarking passengers, but prior to boarding as well. Advise them to consider every prop "hot" at all times.

FAR 91.107 (a) reads in part, "No pilot may take off in a U.S-registered civil aircraft...unless the pilot in command of that aircraft ensures that each person on board is *briefed* (emphasis added) on how to fasten and unfasten that person's safety belt and shoulder harness, if installed. The pilot in command shall ensure that all persons on board have been notified to fasten their safety belt and shoulder harness, if installed, before takeoff or landing."

The repeated reference to "shoulder harness, if installed" is based on the fact that it is only fairly recently that newly manufactured aircraft must be equipped with shoulder restraints. And it was not mandated that aircraft built before this requirement be retrofitted with shoulder harnesses. In fact, in some earlier airplanes it is not possible to install them.

The regulation goes on to require, (b) "During the takeoff and landing of a U.S.-registered civil aircraft...each person on board that aircraft must occupy an approved seat or berth with a safety belt and shoulder harness, if installed, properly secured about that person. However,... a person on board for the purpose of sport parachuting may use the floor of the aircraft as a seat." The regulation doesn't say, but it seems to me that it is implied that if the seats have been removed for skydivers to sit on the floor, the seat belts which are attached to the floor should be fastened about the thighs of those sitting on the floor. (See Chapter 1 for what can happen if they are not.) I got fired by a sport parachuting club one time from a job flying skydivers because I insisted on those guys sitting there on the floor being restrained with belts. They bitterly resented it, but I refused to take off until they were all secured. Somehow they seemed to think that since they didn't plan on sticking around for the landing, they weren't required to wear seat belts for takeoff.

And while we're on the subject of passenger briefing, how many of you make sure your passengers know how to get out of the airplane if you should have a mishap in which the airplane ends up on the ground inverted with the door(s) jammed? Not many, I'll wager. Yet this is something that is supposed to be included in the passenger briefing. Just as the cabin attendant in an air-carrier aircraft briefs the passengers on evacuation, so should you. Do you ask your passengers to help watch for traffic? It's not regulatory, but it is a good idea, don't you think? What else should be included in the passenger briefing? You tell me.

And while we're on the subject of responsibility, I cannot emphasize strongly enough that the ultimate responsibility rests on you, the pilot. Many pilots, even fairly high-time pilots, are prone to relax and leave it all up to ATC when they are in radar contact. Do not do this! It is up to you to assure your own safety. Occasionally you may even have to demand your right to do something (or your right to refuse to do something you believe to be dangerous).

Pilots who run out of gas (and lots of 'em do) have nobody and nothing to blame but themselves. This kind of activity clearly demonstrates a lack of personal responsibility. Have you ever permitted yourself to become fuel critical? Perhaps there were unexpected headwinds, but whatever the reason, we cannot permit this to happen. There are surely enough airports wherever you're going (unless its across the big puddle) for a fuel stop if the headwinds are so strong as to prevent you from reaching the planned stop.

I simply hate to keep reading accident reports where the probable cause boils down to "pilot error." Lately, we've been hearing a lot about CFIT (controlled flight into terrain), and this is another one for which there is no excuse. It can only happen when pilots leave their protected airspace, another clear demonstration of the lack of personal responsibility.

# II Checklists

Another area of responsibility in which many pilots fail to do an adequate job is in the use of checklists. Some pilots don't use the checklist at all, while others run through it in a perfunctory manner, paying only lip service to its importance. As an example of this, let me relate a story told to me by an experienced flight instructor: it seems that one of the airplanes belonging to the school where he worked had an inoperative oil pressure gage. The needle just lay dead at the left end of the gage. No less than eight students in a row failed to notice this. With the checklist in hand, they would start the engine, reach out and touch the face of the gage and say, "Oil pressure coming up."

At the runup pad, they would again run through the list, confirming that all the engine instruments were "in the green." The instructor would then ask the students to do it again, and they would finally notice that the oil pressure gage was reading "zero." I would really like to have seen the expressions on the faces of these students when they realized just what they had missed the first two times through.

As another example, let me cite a technique I used when giving a private pilot practical test. After the walkaround preflight inspection when the applicant and I would board the airplane, I would lay the seat belt across my lap, leaving it unfastened, or I would simply sit on top of the lap belt. Almost invariably the applicant would fail to notice this. When he/she had completed the engine runup, the applicant would reach for the mic to call ready for takeoff, at which point I would stop him or her and ask, "Are you ready?"

*Applicant: Sure am.*

*Me: You sure you're ready?*

*Applicant: Yup.*

*Me: Run the checklist, did you?*

*Applicant: Sure did.*

*Me: Do it again.*

The applicant would then go through the list, usually starting with the pretakeoff section. Most checklists start off with item one of the prestart list, something like "Belts and harnesses fastened and seats adjusted." Then on the pretakeoff portion of the list, they repeat the same item. Even so, most applicants would still miss it the second time through, and the conversation would continue:

*Me: You sure you covered everything on the list?*

*Applicant: Sure am.*

*Me: Go back and check item one.*

And many of them would still miss it!

What these examples tend to demonstrate is that many if not most pilots don't really use their checklists. They just kind of hurriedly brush over the items on the list. On the checklist of every single airplane with a carbureted engine, there is an item regarding "check carb heat" during the runup at 1,800 or 2,000 rpm, or whatever. Most pilots yank the carb heat knob out (one once pulled it out so hard the cable came right out in his hand), note an appropriate rpm drop, and shove it right back in, failing to check it at idle. If you want to be sure your engine will respond when you most need it to, you'd better check the carb heat at idle rpm and confirm that it continues to run okay. The reference to "most need it to" above is in the event of a last-minute go-around, when the engine is idling, just before the expected touchdown.

One of my special heroes in aviation, Pete Campbell, formerly with the FAA, and beyond doubt the greatest aviation speaker I've ever had the pleasure of hearing, has always emphasized the passenger briefing item on the checklist. And the rest of the world is finally beginning to catch up to him with the recent emphasis on CRM, or cockpit resource management. No matter who else is aboard (anybody from a senior airline captain to a small child), it is imperative that on every single flight the person acting as PIC thoroughly brief passengers on just what to expect and when, from instructing them on the use of belts and harnesses to the technique for escaping from the wreckage after the crash.

And how about the rest of the paperwork? Pilots know they must have their pilot certificates and medical with them when exercising their pilot privileges, and pilots also know that the airplane is required to have ARROW (airworthiness certificate, registration, radio station license, operating limitations, and weight and balance data) aboard, but do pilots really know and understand each of these items?

How is the airworthiness certificate renewed? Really, there are three answers to that question. One, it is never renewed in the sense that a new certificate is issued. It is originally issued under the authority of the FAA by the manufacturer when the airplane is built and released for service, and it stays with the airplane throughout its lifetime. Two, it is renewed by the mechanic with inspection authorization when the mechanic signs off the annual inspection (or by the mechanic who signs off the 100-hour inspection) by the statement that "I have personally inspected this airplane and find it to be in airworthy condition," with the total time in service, dated and signed with the mechanic's certificate number. Once it is signed off like that, we know it was airworthy on the date entered and with the experience (number of hours) it had at that date and time, but how do we know it is airworthy today? (A few days and a few more hours of experience later.) However, the third and best answer to the question of how an airworthiness certificate is renewed is number three. When the pilot in command completes the walkaround preflight inspection, does the pretakeoff runup, and applies power to take off, the pilot is stating, in effect, "I have personally inspected this airplane and find it to be in airworthy condition for the operation I intend to perform today!"

# III Registration

How about the aircraft registration? Is it for real, or is the expired temporary registration the only thing aboard?

I have known several instances where a new owner put the temporary (pink slip) registration in his airplane, and sometime within the next 90 days when he received the permanent registration, he neglected to replace the temporary one in the airplane with the real one. In this regard, an interesting thing happened to me a few years ago. A guy who had bought a brand new Skyhawk and leased it back to the dealer from whom he had bought it decided to change facilities and lease it to my flight school. It had gone through no less than eight 100-hour inspections at the original operator's school and five at my school when I rented the airplane to a brand new pilot, a young lady who had passed her private pilot flight test the preceding day. She took a friend and flew some 60-odd miles away to another airport for lunch. When they arrived at the distant airport, disembarked, and walked into the terminal where the restaurant is located, they were met by an FAA inspector, and the following conversation occurred:

> *Inspector: Hi! I'm (gave his name and produced proper identification). May I please see your pilot certificate and medical?*

> *Pilot: Of course. (proudly shows both).*

> *Inspector: Airplane paperwork all in order? (He knew me and my school's reputation for always being in full compliance, and he had seen my name on the temporary certificate issued the previous day.)*

> *Pilot: Sure is! Come on, I'll show you. (Big mistake! He would have forgone this if she hadn't insisted.)*

When they walked out to the airplane and looked at all the paperwork therein, they found everything else in order, but the only registration aboard was a temporary one that had expired some 10 months previously.

They then went back into the terminal and called me up to inform me of the situation. I explained the history of the airplane to the inspector and readily admitted that I had made a stupid assumption (you know about assume making an ass out of u and me) in believing that a reputable outfit such as the one that had the airplane before we got it would permit it to go through all those inspections without catching the fact that the permanent registration had not replaced the temporary one in all that time, let alone the time we had it and the inspections our own shop had performed on the airplane. I also explained that the owner, a nonpilot, had no doubt received the permanent registration and thrown it into a desk drawer at his home. I told him that I would fly over to the location where my renter and her friend were and bring them back, leaving the airplane without a valid registration there until we could obtain the permanent one from the owner. Being a really nice guy, he said, "Howard, I'm leaving right now. I'm going to walk out the front of this terminal, get in the G-car in which I came and drive away." And then to make sure I got it, he repeated this statement. The implication was, of course, that the two women could have their lunch and fly the airplane back and nobody would be the wiser. Even so, I did not accept his kind offer. I flew over there, picked up the two girls and brought them back. After I advised the owner to look for the permanent registration, he found it and dropped it off the next day, after which we retrieved his airplane, all this at his expense. Quite a lesson, wasn't it? And there are several other paperwork items about which the pilot should be particularly knowledgeable and frequently isn't, such as the *Approved Flight Manual.*

# IV Approved flight manual and equipment list

The basic rule holds that for each individual airplane built, there shall be an approved flight manual, setting forth the operating limitations and including an equipment list with sufficient data to work a weight and balance. The regulation goes on to state that the manufacturer of an airplane with a maximum certificated gross takeoff weight of less than 6,000 pounds may, in its option, publish the operating limitations in the form of placards and markings placed about the cockpit in plain view of the pilot. If this option is exercised, the manufacturer must also include a separate equipment list with sufficient data to work a weight and balance.

Can you guess how many pilots mistakenly believe the owners handbook is an approved flight manual? Most of 'em, that's how many! And how many do you suppose believe it is okay to use the sample weight and balance data in the owner's handbook instead of the actual figures from the equipment list (FAA Form 337—Major Alteration and Repair form, or other official and accurate data)? A whole bunch. That's how many! Much of this kind of confusion got cleared up when, in 1978 the General Aviation Manufacturer's Association (GAMA) got the airplane manufacturers to agree to quit exercising the option, and they all started supplying their airplanes with approved flight manuals. Even so, some of these approved flight manuals look just like the owners' handbooks and a certain amount of confusion still exists. (The real thing is in looseleaf form so it can be revised, while the owner's manual is usually in bound form.) And a substantial amount of flight training is done in airplanes which were built before the change and still use the placard technique of delineating the operating limitations of the airplane.

And if an item is on the equipment list, it must be in or on the airplane. You want nitpicking? Here's one for you. Back when Cessna was producing light single-engine airplanes when a new airplane (150, 152, or 172) was shipped from the factory, there was included a cargo net in a cellophane envelope about 9 or 10 inches square and 1½ inches thick. This item was on the equipment list. However, most dealers would remove this little packet and throw it in a bin somewhere before putting the airplane in service. One day a student took such a CE 150 (without the cargo net) to the local FAA District Office for a checkride. Noting that the equipment list included a cargo net and the airplane didn't have one aboard, the FAA safety inspector (General Aviation Operations type) grounded the airplane as being unairworthy and sent the applicant back with a ferry permit!

How's that for nitpicking? I've heard of similar cases in which an airplane was deemed to be unairworthy because it didn't have a tow bar aboard, and a tow bar appeared on the equipment list. I've even heard of the FAA declaring an airplane to be unairworthy because the cigarette lighter (which was on the equipment list) was missing from the airplane. Of course, you may have to look in several places to find all the stuff that belongs on the list. As an airplane grows older, various items (particularly avionics) are removed, replaced, deleted, and added, and such information may be entered in the aircraft log or on a 337 (Major Alteration or Repair) Form setting forth the change in the weight and balance information for that particular airplane.

I'm sure you already know how to tell if you are using the most recent (current) data for figuring the weight and balance and center of gravity location. You say you don't? Simple. It is the one that doesn't say "superseded," because every time there is a change in the weight or center of gravity location, the individual who makes the change and entry in the aircraft records is required to write "superseded" on the last entry (wherever it is), and the date and location of the new one. By following this trail you can find the current data.

# V  Service bulletins and airworthiness directives

After a specific design has been tested and approved, the manufacturer is issued a type certificate, and production of that model of airplane can begin. Regardless of the amount and depth of testing done, after that model finds its way into the marketplace and gains some experience in the hands of the flying public, certain factors come to the attention of the manufacturer and/or the FAA, and it becomes apparent that certain safety changes must be made in future production and in those individuals of that model already in the hands of the public.

Frequently, but not always, the change comes about first in the form of a service bulletin (SB) issued by the manufacturer. These bulletins are recommendations only and not mandatory with respect to compliance. Again, usually, but not always, the SB is followed by an airworthiness directive (AD note) issued by the FAA. This is a mandatory order to fix or change whatever is required to render the airplane safe. It may require a repetitive inspection, the replacement of a component, immediate action, action at a specific time in service, or action by a specific time after issuance. It behooves the owner of an airplane to be familiar with all the service bulletins and airworthiness directives that affect his or her airplane, and see that the airplane is in compliance for the ultimate responsibility falls on the owner/operator.

# VI The minimum equipment list

The rule governing MELs, or minimum equipment lists, has been with us forever and has always been applied to air-carrier aircraft, but only recently has the FAA been paying attention to this regarding general-aviation airplanes, and even now, only really with regard to light twins. For the most part it is still being ignored insofar as single-engine airplanes are concerned. But this is not always the case, so you'd better know and understand just what it is and how it works.

Basically, as pointed out above, if it's on the equipment list, it had better be on the aircraft. If its on the list, not only must it be in or on the airplane, but it has to be working! As mentioned previously, I know of a case where the FAA grounded an airplane as "unairworthy" because the cigarette lighter wasn't there! The socket was, but the lighter element was not. More nitpicking, huh?

If a nonessential item (instrument or radio for instance, or any item not required by the type certificate of the airplane or by the flight being conducted) is not working, it must be physically removed from the airplane, and the removal noted in the aircraft records. In the alternative, it may be placarded as "inop." However, since there are a great many things in and on every airplane (cigarette lighter, for instance) that are not required for flight, owners or operators may construct minimum equipment lists for their specific airplanes. If they do this, the list must be approved by the FAA, and this entails a ping-pong game of having the suggested list bounce back and forth between the owner/operator and the FAA until final approval is granted.

The FARs spells out what is required for day VFR, night VFR, and IFR flight, but virtually nobody climbs aboard an airplane with the intention of flying with only the legal minimum stuff. We all want everything we can afford and that will fit in the airplane. I know I do. How about you?

I could go on forever explaining how and why it is important for pilots to be thoroughly familiar and careful and accurate regarding their paperwork and that of the airplanes they fly, but by now I'm sure you've got the idea.

# VII The private certificate

There is always confusion regarding the ability of a private pilot to share the expenses of airplane operation with passengers. In fact, the rules are so stringent and interpreted so harshly that, with the exception of politicians running for office, private pilots would be better off if they just absorbed all the cost of whatever flying they do. Politicians! Boy, those guys really know how to take care of themselves, don't they?

You say you'd like to do something good for God, country, and motherhood? For your fellow humans? Be careful. Don't rush into anything. Take a good, long look at the consequences. If you're a pilot and value your pilot certificate, you may want to look twice before leaping into the area of flying for a charitable foundation. And after taking a long, hard look, you may want to back off and permit some poor sick individual to die alone, lacking the medical care he or she could get if you and your airplane were available.

You see, the friendly feds have a strange way of interpreting the regulations they write. It seems that if a charitable organization is eligible for a tax exemption, even if pilots don't claim a deduction for out-of-pocket expenses, any flights they make for the charitable outfits are considered commercial operations, and the pilots could be held to be in violation of FAR Part 135 unless both the aircraft and the pilots were on a 135 Air Carrier Operating Certificate. The mere fact that they're eligible for such a deduction makes them guilty of operating in violation of Part 135 of the Federal Aviation Regulations. Bear in mind the pilots didn't take the tax deduction for which they were eligible. But the fact that they were entitled to a deduction was sufficient for the FAA to call it flying for compensation or hire and thus a commercial operation. This obvious abuse of the power granted to the agency by the U.S. Congress is just one example of the arrogance with which the management of the FAA operates. And, of course, private pilots who take or are entitled to take tax deductions for the expenses involved in charity flights are deemed to be guilty of commercial operations in violation of the privileges of their Private Pilot Certificates.

This situation arose when a private pilot who frequently rented a Beechcraft Baron for his business use was asked to pick up and deliver a passenger for a nonprofit, tax-exempt organization. This he did, and although under the rules of the IRS he was entitled to take a tax deduction for the rental of the airplane, he did not do so. When the flight became known to the FAA, the pilot was charged with violations of several FARs (violation of Part 61 by operating an aircraft for compensation or hire when he had only a private certificate, and acting as pilot in command of an aircraft in a commercial operation without the appropriate Air Carrier Certificate under Part 135). The FAA sought to revoke his pilot certificate, but on appeal, the NTSB reduced the penalty to a 30-day suspension.

The NTSB appeal opinion says, in part:

> ...the law judge found that the respondent had...operated an aircraft for compensation or hire when he did not possess necessary commercial operating authority. He therefore affirmed an emergency order of the Administrator in the extent it alleged that respondent had violated sections 61.118 and 135.5 of the Federal Aviation Regulations, "FAR," 14 CFR Parts 61 and 135. However, the law judge, finding no precedent or justification for the sanction of revocation sought by the Administrator, modified the order to provide for a 180-day suspension of the respondent's private pilot certificate.

Suddenly pilots—both those with commercial pilot certificates and those holding only private pilot certificates, who have for years been donating their time, talent, and airplanes to fly for charitable organizations—are becoming aware of just how far the FAA will reach in an effort to curtail aviation by violating pilots and are becoming terrified of losing their privileges. And this is the agency that is charged with the responsibility of promoting the development and growth of civil aviation in the United States.

We have a unique program in our district which has a substantial number of pilots with aircraft playing Santa Claus each year at Christmas time. A huge collection of toys and clothes (enough to fill a large hangar) is gathered from the donations of generous citizens. These gifts are then flown to destinations throughout the state for distribution to children who otherwise would suffer a bleak Christmas. The pilots and aircraft owners receive no compensation (not a dime of expense money) for this service. Now—since the outrageous interpretation of the regulations that hold that since they are eligible for a tax break by virtue of donating their time, talent, and equipment to this worthy cause, they are in violation of one or more of the Federal Air Regulations—several of the pilots with whom I've spoken will no longer participate in this program.

Another group of pilots and aircraft owners who are reconsidering a worthwhile activity in which they have been engaged for many years are the people involved in the "Mercy Med-Flights." Locally, this is part of a nationwide network of pilots who own or rent airplanes on a volunteer basis for the transportation of organs, plasma, specimens, and people who would otherwise be unable to reach an appropriate destination for medical purposes. All these aviators are ceasing or considering ceasing the wonderful activity in which they are engaged because of the unreasonable interpretation the FAA places on its own regulations, all in the name of safety.

In stark contrast to this ridiculous interpretation of one of their own regulations is the special treatment accorded a specific class of special privileged citizens. In an obvious attempt to placate the Congress, who holds the purse strings of the agency, the FAA enacted specific regulations exempting the transportation for hire of politicians running for federal office from complying with Part 135 of the regulations. Thus any office holder, would-be office holder, staff member of an office holder, or staff member of a would-be office holder can hire any pilot with an airplane (or helicopter, airship, balloon, or glider) to fly them anywhere so long as they are campaigning for office (or in the case of a staff member, working on the campaign of a candidate for federal office). It seems that our government really takes care of its own!

The regulation covering this disgraceful activity is Part 91.321, and it reads as follows:

> (a)*An aircraft operator, other than one operating an aircraft under the rules of Part 121, 125, or 135 of this chapter may receive payment for the carriage of a candidate in a Federal election, an agent of the candidate, or a person traveling on behalf of the candidate, if—*
>
> > (1) *The operator's primary business is not as an air carrier or commercial operator;*
> >
> > (2) *The carriage is conducted under the rules of this Part 91; and*
> >
> > (3) *The payment for the carriage is required, and does not exceed the amount required to be paid, by regulations of the Federal Election Commission (13 CFR et seq.).*
>
> (b) *For the purposes of this section, the terms "candidate" and "election" have the same meaning as that set forth in the regulations of the Federal Election Commission.*

The United States of America is still the greatest country in the world, and I value my citizenship here very highly. However, our government is not without flaws as this disgusting display of self-service by those who are elected and hired to serve the public illustrates so well. The arrogance of some bureaucrats when granted a little power has enraged a huge number of its citizens almost to the point of rebellion.

# VIII The commercial certificate

It never ceases to amaze me how commercial pilots, who seem to be among the majority, will demonstrate a complete lack of understanding regarding just what they may or may not do to earn money as pilots. As a pilot examiner conducting a practical test for the commercial pilot certificate, I would always ask, on the oral portion of the test, just what the applicant could do to earn money as a commercial pilot. Demonstrating an appalling lack of understanding of the difference between Part 91 and Part 135 of the regulations (which requires a "Holding out" and "Operational Control," both of which are subject to the unique interpretation of the FAA), I would usually get either, "All I can do is give air rides within 25 miles of the departure airport under day VFR, or, if I get an instructor certificate I can teach," or, "I can do anything by way of flying for hire!" I would then set up a hypothetical and quiz the applicant as follows:

"Your friend Bill's wife, Nancy, calls and informs you that Bill was in an automobile accident in a distant city. He is in the hospital and he wants me there as soon as possible, and he asked me to call you and ask if you'll fly me there in your airplane (either rented or owned). I'll be happy to pay you for taking me there. Will you do it?"

I would then ask the applicants if they could legally do this and get paid for it. The answer, of course, is no. Since you own (or control) the airplane, you are exercising operational control. You are supplying an airplane and a pilot.

I would then set up the same situation, but this time Nancy is asking you to fly her in Bill's airplane, and again she's offering to pay you for your services as a pilot. In this case it is perfectly okay. You are providing commercial pilot service in an aircraft owned or controlled by the customer.

Many commercial pilots don't realize that they can be employed as professional corporate pilots carrying the company's equipment, personnel, and guests. However, if the guests pay all or even part of the expenses of the trip, then the corporation for which the pilot works would have to have a Part 135 Operating Certificate, and the pilot would have to meet 135 requirements and be named under the certificate. If all this sounds complicated, it is. But it is stuff that commercial pilots must know if they are to avoid violating the regulations by inadvertence.

# IX The instrument rating

In almost every written test for the instrument rating, the FAA forcefully brings to the attention of the applicant one or both of two very dangerous privileges bestowed upon those who hold the rating.

The first and really dangerous one is the fact that the regulations require that in order to operate under IFR in IMC (instrument meteorological conditions) in controlled airspace, the airplane must be properly equipped and the pilot appropriately rated, and on a flight plan with a clearance. There are two bodies of law, those which are restrictive in nature (if it doesn't say you can't do it, you can), and those which are permissive in nature (you can't do anything unless it is specifically authorized). Since the FARs are restrictive in nature (They all start out, "No person may...unless..."), this means that an instrument-rated pilot flying a properly equipped airplane may blunder around in uncontrolled airspace without being on a flight plan and operating with a clearance from ATC. Stupid? You bet it is! The purpose, of course, is to enable pilots to take off from an uncontrolled airport in uncontrolled airspace, expecting to be on a clearance as soon as they enter controlled airspace, and this is a valid purpose. But the regulation could certainly be better expressed. As it now stands, if pilots could stay out of controlled airspace (perhaps by remaining under the floor), just so long as they are rated and equipped, they could go anywhere and do anything. In other words, it is perfectly okay to go blundering around in the cloud where some other idiot might be blundering around just so long as you do it in uncontrolled airspace. And although there is very little of it left, there are still some places where uncontrolled airspace can be found above 1,200 feet above ground level.

The other area, which isn't nearly as bad, but which has the potential of tying up a huge amount of airspace, is based on the requirement that under IFR the pilot must have navigation equipment appropriate to the ground facilities being used. (This is being changed to accommodate GPS, which, of course isn't based on "ground facilities," but on satellites.) What this regulation means is that in a radar environment, all pilots are required to have to go from point A to point B is a transponder and a two-way communication radio. They need only put in the "remarks" section of their flight plans, "Radar vectors only and radar approach only." How about that? Of course, they can expect to be waltzed around a few times as the controllers attempt to keep them in radar contact, but they may still legally do this.

# 13

# What do you think of the FAA?

## I  ATC and flight standards

The title of this chapter is taken from an area in the FLYING forum on America Online. I've been reading and occasionally contributing to this area for quite some time, and I will herewith pass on some of the interesting stuff I have learned.

I have made no secret of the fact that I've had a more than 50-year love affair going with ATC. I really believe those guys in the Centers, TRACONS, RAPCONS, Air Traffic Control Towers, and flight service stations are the greatest public servants in the history of the human race. For the most part, these people treat us, the users of the airspace, like customers, which, of course, is just what we are. As is the case with most large government organizations, the collective attitude of the staff filters down from the top, and the attitude of the bottom-line people is a reflection of that of the facility manager. I am aware of at least one ATCT (Air Traffic Control Tower) in which the manager has his people keep small problems in-house. If a pilot by inadvertence blunders into committing some minor violation, endangering nobody, it is not reported to Flight Standards for processing as a violation, but rather handled locally by means of counseling.

However, the Flight Standards division of the FAA is another story entirely. Back in Chapter 12 I related a couple of horror stories regarding the zealousness with which some aviation safety inspectors interpret the regulations and the apparent glee with which they issue violations against errant pilots and operators. Shortly after TCAs (terminal control areas) were set up at the busiest air carrier airports in the country, on a Monday morning I had occasion to visit my local FSDO, and one of the general aviation operations inspectors was happily shuffling papers around on his desk. It was the only time I'd

ever seen this guy smile. When I asked him why he seemed so joyful, he replied that he really liked Mondays because that's when he
got to work all the TCA violations that occurred over the weekend.
How's that for a viscous attitude?

The thing to remember is that way back when the Congress of the
United States created the CAA (Civil Aeronautics Authority), predecessor of the FAA, they charged the agency with the responsibility
(read "duty") to "foster and encourage the development and growth
of civil aviation in the United States." Way down the list it was authorized (read "allowed") to promulgate and enforce regulations
pertaining to safety in air commerce. Over the years since that
founding of the agency, it has moved further and further away from
encouraging and fostering the development and growth until now
almost the entire energy of the agency is devoted to promulgating
and enforcing regulations, many of which have absolutely nothing
to do with safety. The claim is often made that a pilot has endangered the life or property of another when there is nobody anywhere around to endanger. Back in Chapter 6 I cited a couple of
typical horror stories to illustrate how far the FAA would reach in
stretching the regulations to nail a pilot with a violation.

As a direct result of this activity on the part of the FAA, an adversarial position has arisen between the aviation community and the
agency. Because of the high-handed and devious manner in which
the agency traps pilots and operators into admitting wrongdoing, we
have reached a point at which virtually nobody in the aviation community trusts the FAA, and that's too bad because it should be a part-
nership between the providers and the users. Fortunately there are a
few—a mere handful of aviation safety inspectors—who view their
public service jobs as actually existing to serve the flying public, but
these good guys seem to be swimming upstream. It appears that
they are constantly fighting a losing battle against a system that re-
wards mediocrity and punishes those who try to do a good job.

The old joke about, "Hi, I'm from the FAA and I'm here to help you."
may be funny, but it is also tragic in its implications. Back in Chapter 6, and again in Chapter 12, I cited examples of how far they can
and do stretch the interpretation of the regulations they write and
enforce. A correspondent of mine wrote and said, "I love what the
FAA does for us in the control towers and on the airways, but I hate
what their offices in Washington and Oklahoma City have become.
The FAA is not your friend."

He goes on to relate how an operator in the district where he lives and works recently got cited by the friendly feds for not having proper de-icing equipment and procedures in place. This operator, who conducts sightseeing flights in seaplanes around the Keys in South Florida, was cited for failure to have de-icing equipment and procedures. Their floatplanes never operate above 1,500 feet above the ocean, where the temperature never drops below about 78 degrees Fahrenheit. These are day VFR-only sightseeing flights, but the FSDO (Flight Standards District Office of the FAA) doesn't care. They say the operation needs to address the subject of airframe ice. My correspondent adds, "The only icing problem these guys will ever have is when the ice in their drinks melts too fast."

This is in the same category of silliness as grounding an airplane as unairworthy (read "unsafe") because it didn't have a cargo net or towbar aboard, or cigarette lighter aboard when one or another of these items was on the equipment list (see Chapter 12). Just how unsafe is it to fly without a cargo net if you're carrying no cargo, or to operate in the tropics without anti-ice or de-ice equipment? It may seem silly to you or me, but we must remember that many of the bureaucrats in the FAA are totally devoid of a sense of humor and completely out of touch with the real world.

Just how far will they go to nail a pilot or operator with one of these phony violations? I don't know, but I have heard of cases where airplanes were grounded as unairworthy, that is "unsafe," because there was no cigarette lighter in the receptacle on the panel, although the equipment list included a cigarette lighter. I question just how accurate these particular reports are, but anything is possible. I do know from personal experience that the FAA will go as far as necessary when they are after a specific individual. On one occasion they took the unheard-of measure of sending an inspector out of his district, even out of his region, with specific instructions to ground an airplane of mine which was parked on the ramp at a distant airport. After three days of crawling all over my airplane, he grounded it because, get this, it had a broken static wick! And this is an airplane that isn't even required to have a static wick.

Have you ever taken off without having a current chart aboard? Have you ever blundered into airspace where you didn't belong? Have you ever inadvertently busted an assigned altitude? Have you ever let the date for your flight review or medical exam slip by without having noticed? Because there are so many regulations, many of which have no relationship with safety, it is almost impossible to fly for any length of time without committing some sort of violation. Fortunately, most of these minor violations never come to the attention of flight standards, but those that do invariably cause the pilot no end of trouble.

# II The letter of investigation

Why should pilots, mechanics, and operators trust the agency when they have been consistently lied to and deceived by inspectors? It all starts when the alleged violator receives a registered letter. The infamous LOI (letter of investigation) is a perfect case in point with respect to the kind of deception practiced by the FAA. It is the aviator's invitation to hang himself or herself. It is cleverly worded so as to appear to require an answer, but it is not mandatory that an answer be offered. It goes like this: (paraphrasing) It has come to the attention of this office that a violation of... (section number) of the Federal Aviation Regulations may have occurred on... (date)... at ...(location), and we have reason to believe you may have been involved. We would appreciate receiving any evidence or statements you might care to make regarding this matter. Any discussion or written statements furnished by you will be given consideration in our investigation. You have 10 days to respond to this letter of investigation or we will be forced to proceed without benefit of hearing your side of the story.

This is your invitation to hang yourself. If you answer and admit you busted the assigned altitude (or whatever violation is being investigated) and offer what seems to you to be a reasonable excuse, you will no doubt be violated and suffer a certificate suspension or worse. And since this is an administrative civil action the courts have ruled that although there's no warning to that effect, there is no Fifth Amendment protection against self-incrimination. In plain English, the United States Constitution doesn't apply to the FAA! Also, please note that there is no mention of the possibility of remedial training in lieu of a sanction (certificate action or civil penalty).

If you should ever be so unfortunate to receive one of these dreaded LOIs, immediately retain the services of an attorney who is familiar with the administrative procedures of the FAA. The attorney will either advise you not to answer the LOI, will reply in your behalf, or will tell you to answer but admit nothing. For example, you might say, "Thank you very much for the opportunity, but I do not wish to make a statement at this time. Please direct further correspondence to my attorney, Mr./Ms...." Nobody likes to be ignored, and if you address such a letter as that to the specific inspector who signed the LOI, at least you didn't ignore him or her.

My point is that the sneaky way the LOI is worded leads average pilots reading it to believe that they are required to answer and spill their guts. And although there is no warning to that effect, anything you say can and will be used against you at either an informal hearing (a meeting with an FAA attorney from the regional office), or a more formal hearing before a NTSB (National Transportation Safety Board) administrative law judge, at which time the FAA attorney, who has been both prosecutor and judge at the informal level, now takes on the role of prosecutor before the ALJ (administrative law judge). This is very similar to a criminal case in a law court. Sworn testimony is taken and a record is made by a court stenographer.

# III  The violation procedure

The next step after the LOI is another letter, also by certified mail (return receipt requested) entitled notice of proposed certificate action. This letter states that at the conclusion of the investigation referred to in the LOI, a finding of violation has been made. It cites the specific regulation or regulations which are alleged to have been violated and the specific actions which constituted the violations, along with the date and place of occurrence.

This "notice of proposed certificate action" goes on to state the penalty the FAA proposes to impose and it concludes by offering the airman five choices. (Just check the appropriate box and return the letter to the Office of the Regional Counsel.)

It says, and I quote:

> *In reply to your Notice of Proposed Certificate Action..., I elect to proceed as indicated below:*
>
> *1. I hereby transmit my certificate with the understanding that an Order will be issued as proposed effective the date of mailing of this reply.*
>
> *2. I request that the Order be issued so that I may appeal directly to the National Transportation Safety Board.*

*3. I hereby submit my answer to your Notice and request that my answer and any information attached thereto be considered in connection with the allegations set forth in your Notice.*

*4. I hereby request to discuss this matter informally with an attorney at the Office of the Regional Counsel...*

*5. I hereby submit evidence of the timely filing of an Aviation Safety Report with NASA concerning the incident set forth in the Notice of Proposed Certificate Action and thereby claim entitlement to waiver of any penalty.*

Almost nobody checks the first box and surrenders his or her certificate. The majority of airmen (and the term includes women as well) opt for choice number four, although many lawyers advise their airman clients to skip this step and go directly to the NTSB by selecting choice number 2, which results in a hearing before an NTSB administrative law judge (ALJ).

If number four is selected, as is most often the case, a so-called informal meeting is held at the alleged violator's local Flight Standards District Office (FSDO). Present, and sort of moderating this meeting, will be an FAA attorney from the Office of the Associate General Counsel, formerly the Office of the Regional Counsel, the local FAA aviation safety inspector who conducted the investigation, who acts as a sort of prosecutor, the alleged violator, and if he or she has one, an attorney.

After a full discussion of all the facts surrounding the alleged violation, the FAA attorney may close the matter with a finding of no violation (extremely unlikely), modify the penalty (sanction) by, for example, knocking down a proposed 60-day suspension to 30 days (very unlikely), or affirm the sanction as proposed. In many instances the attorney has no authority to do anything. The attorney is sent from the regional office with instructions to impose the proposed penalty and that's all the attorney can do.

Whatever the outcome of this informal meeting, the airman may appeal to the NTSB, just as if he/she had chosen option number 2 in the Notice of Proposed Certificate Action.

At this point a formal hearing before an ALJ is scheduled. It is not really scheduled in terms of a specific date, but it is put on the agenda. Sometime in the future (quite a long time), a date for the hearing is set. At this hearing sworn testimony is taken, and a record is made (court stenographer and all).

The attorney from the Office of the Regional Counsel who ran the informal meeting now takes the part of prosecutor, and the hearing is similar to a trial in a law court. The investigating inspector takes the stand and is questioned by the FAA attorney and cross examined by the alleged violator's lawyer. Other witnesses may be called to give testimony, both the FAA's and the airman's witnesses. This hearing takes on the appearance and nature of an actual trial in a law court.

Again, at this level the ALJ may confirm, or reverse the decision to violate the airman, or the ALJ may modify the penalty (reduce the suspension from 90 days to 30 days, for instance). If either the airman or the FAA don't like the decision and have a valid reason to do so, either of them may appeal to the NTSB itself, and once again after a lengthy wait—often many months, the board may affirm or reverse the decision of the ALJ, or modify the penalty, if any.

If an appeal is taken to the full NTSB, the hearing at that level is more like that at a court of appeals in which material previously presented is reviewed, but additional evidence may also be considered. Again, at this level, the decision of the ALJ may be affirmed, modified, or overturned.

Unless the FAA has exercised its power of emergency certificate revocation, the airman keeps his or her certificate and continues to exercise the privileges of the certificate throughout the entire process. And this can take many months, or even years.

## Revoke!

One of the most blatant abuses of power at the hands of the FAA has been with the emergency revocation power. The exercise of this technique was designed to prevent the continuation of an unsafe condition or activity, but the agency hasn't hesitated to use it in cases where no emergency whatever existed, almost to the point that it has become routine. Fortunately for the aviation community, an airman who has had his or her certificate revoked under the emergency authority where no emergency exists, if he or she is willing to fight the revocation in the courts, he/she can prevail. Until the case of *Daryl R. Frank et al. v. James B. Busey*, FAA Administrator, case No. 91-1469

(U.S. Court of Appeals for the District of Columbia, October, 1991), the courts have been most reluctant to impose any limits on the discretionary powers of the agency. However, in that case, the court determined that since the investigation had been ongoing for 18 months and the alleged violations had occurred some 18 months previously, there was no immanent danger to safety that would require the emergency revocation of the five pilot certificates in question.

We must bear in mind, however, that this decision hasn't prevented the agency from continuing to abuse its emergency power to revoke the certificates of airmen. It simply means that each case must be fought individually, and if no genuine emergency involving safety can be shown by the FAA, the airman is likely to prevail. Thus, the stubborn refusal of the FAA to relinquish any of the power it has assumed unto itself (without any justification whatever) has forced each individual airman to fight for his or her rights on an individual basis.

## The ASRS

A pilot who has filed, within 10 days of the incident in question, a so-called NASA Report, and who is qualified for a waiver of penalty may check box number 5 in the Notice of Proposed Certificate Action. Notice, I said an airman who is qualified, for there are several qualifying factors which must be present for the airman to avoid the penalty for a violation. Notice I also said avoid the penalty, for the violation stands on the record—that is, the airman's record. It is only the penalty that is avoided.

Several years ago the National Aeronautics and Space Administration set up at the Ames Research Center at Moffett Field in California the Aviation Safety Reporting System (ASRS). This is a system designed to gather data and build a statistical base of incidents which it is hoped will provide a basis for reducing aviation accidents. In order to encourage airmen and women to voluntarily participate and to report hazardous incidents, anonymity is guaranteed those who file reports, and waiver of penalty is offered to qualified individuals who file reports which involve self-incrimination regarding violations.

The conditions under which a violator qualifies for the waiver of penalty are set forth in the information which accompanies the Notice of Proposed Certificate Action. This information is as follows:

*If you have filed an Aviation Safety Report with NASA concerning the incident set forth in the Notice of Proposed Certificate Action, you may be entitled to waiver of any penalty. If you claim entitlement to this waiver, you must present satisfactory evidence to the Administrator that you filed a report with NASA within ten days of the incident concerning that incident. You will only be entitled to a waiver if it is found:*

*a) That the alleged violation was inadvertent and not deliberate; and*

*b) That this violation did not involve a criminal offense, or accident, or disclose a lack of competence or qualification to be the holder of a certificate; and*

*c) You have not paid a civil penalty pursuant to Section 901 of the Federal Aviation Act or been found in any prior FAA enforcement action to have committed a violation of the Federal Aviation Act since April 30, 1975.*

*In the event you prove your entitlement to this waiver of penalty, an Order will be issued finding you in violation but imposing no civil penalty or certificate suspension. Your claim of entitlement to waiver of penalty shall constitute your agreement that this Order may be issued without further notice.*

If the airman submits a report on a form provided and obtained from the FAA (a NASA Reporting Form), he or she will receive, by return mail a receipt showing the date the report was received by NASA at Moffett Field. This is his evidence of the timely filing of the report.

If the penalty is waived by virtue of the airman meeting all the above conditions, he or she may not again use this technique for avoiding a penalty for a period of years. And there's another hook. The record of violation on the airman's permanent record at the FAA Aeronautical Center in Oklahoma City is supposed by law to be expunged after a period of five years, but the FAA is not clearing the airman's record as required by law! How do you like them apples?

The National Aeronautics and Space Administration Aviation Safety Reporting System (ASRS) publishes a monthly newsletter which is yours for the asking. All you have to do is write and ask. The address is:

The Office of the NASA
Aviation Safety Reporting System
P.O. Box 189
Moffett Field, CA 94035-0189

The title of the newsletter you are requesting is "Callback," and each month a specific safety factor is featured and ASRS reports sent in by pilots that relate to that specific area are quoted.

There's one more technique by which the FAA can handle a violation. If it is a minor violation, it is inadvertent, the airman has no history of previous violations, and the airman demonstrates a favorable attitude, he or she may be offered a program of remedial training in lieu of some other sanction (certificate action).

# IV The violation alternative

## The program

It is rare indeed that we have the opportunity to observe (or be involved in) a situation in which everybody wins and there are no losers. But such is the case with one program the FAA has come up with. Lord knows I don't always agree with everything the friendly feds do, but the Remedial Training Program is one that I can wholeheartedly endorse. As far as I can determine, the only thing wrong, the only negative factor, in the Remedial Training Program is that it is not being properly taken advantage of. In other words, it is not being used to the extent that it should be. Here is that rare concept—a win-win situation, and in large measure it is being ignored. There still exists within the FAA a large number of aviation safety inspectors who adhere to an extremely simplistic theory of accident prevention. They are aware of the fact that in every aviation accident there is a pilot involved. Ergo, if there were no pilots, there would be no accidents. Thus the direct avenue to preventing all aviation accidents is to ground all the pilots. Of course, this would also result in the FAA being rendered useless and the agency would no doubt ultimately be abolished, but that fact is one step too far removed for some minds to grasp.

Everybody involved in the remedial training program comes out ahead. The pilots are certainly fortunate in that they not only don't suffer a suspension of their certificates, but after two years the incidents are supposed to be expunged from their records (again, supposed to be, but they aren't). The investigating FAA aviation safety inspectors win, unless they're people who enjoy paperwork, because once the program is offered and accepted, they're out of the loop entirely, thus reducing their workload and the amount of hated paperwork they must turn out. The office of the Regional Counsel is a winner since the remedial training alternative leaves them totally uninvolved, and this even reduces the workload of the drastically overworked administrative law judges who work for the National Transportation Safety Board. The only guy (or lady) who has to work in this situation is the safety program manager (SPM), formerly APPM (accident prevention program manager), and before that accident prevention specialist, or APS. And, of course, this is his or her job anyway. Flight instructors also get to work teaching the program laid out by the SPMs, but then that's what they get paid for doing, and chances are they're looking for the opportunity to practice their profession of teaching anyway.

On May 18, 1990 the administrator approved Compliance and Enforcement Bulletin 90-8, Corrective Action Through Remedial Training, and it became effective on that date. By June 12 of that year, just over three weeks later, it was necessary for the Associate Administrator for Regulation and Certification, in concert with the Chief Counsel, to get out a memorandum to all regional flight standards division managers and regional assistant chief counsel explaining and clarifying the bulletin. The purpose of this memo was to prevent the FAA from making procedural mistakes that would permit a violator to escape unpunished.

On October 11, 1990 Notice N 8700.4 was published by the then acting director of the Flight Standards Service, Thomas Accardi, for incorporation into Orders 8700.1 and 8300.10, the *General Aviation Operations Inspector's Handbook* and the *Airworthiness Inspectors Handbook* respectively. These are the "bibles" by which these people live. Bulletin 90-8 spells out just who is eligible for remedial training and how the program is supposed to work. To begin with, only those pilots who inadvertently commit a violation and are not engaged in commercial activity at the time of the alleged violation are eligible. There are some other factors which will disqualify one from participation in the program, but a negative attitude, a deliberate violation, and a violation committed while engaged in commercial activity are the primary prohibiting factors. A history of previous violation weighs heavily in the determination as to whether or not remedial training will be offered, but even that is not absolutely prohibiting. Although intended primarily for first offenders, in some cases remedial training may be offered to a repeat offender if all other factors indicate that he or she is a suitable candidate.

Until the administrator issued Compliance/Enforcement Bulletin 90-8, titled Corrective Action Through Remedial Training, when a violation was brought to the attention of Flight Standards, the matter was assigned to an aviation safety inspector, either operations or airworthiness as the situation warranted. When, during the course of the investigation conducted by the assigned inspector, the name of the suspected violator came up, a letter would be sent with the investigating inspector's signature. This letter, officially called the Letter of Investigation or LOI, is the invitation for the airmen (pilots or mechanics) to hang themselves. It offers the alleged offender an opportunity to tell his or her side of the story, and it gives him or her 10 days to do so. There is no legal requirement that this letter even be answered or otherwise acknowledged! Even so, the receipt of a LOI is a frightening experience, and many airmen feel compelled by the wording of the letter to spill their guts and start spewing forth everything they know or think they know about the situation, and whatever they say not only can but most certainly will be used against them in whatever future action follows. In other words, the LOIs are a basic invitation for the airmen to incriminate themselves, and the letters very often work in just that fashion. Remember, in this situation there is no constitutional protection against self-incrimination.

Compliance/Enforcement Bulletin 90-8 mandates, among other things, that the LOI must include mention of the possibility of remedial training, where appropriate. Without committing the FAA to a firm offer, the LOI is now supposed to include mention of remedial training as an alternative to sanction. The bulletin, in an appendix, even includes a model LOI containing this statement: "Additionally, you may be allowed to participate in the FAA's corrective action through remedial training program, in the place of legal enforcement action that may otherwise be deemed to be appropriate." It then goes on to spell out the requirements for eligibility in the program. Our research indicates that this procedure, mandated by the administrator, is not being followed in some district offices (FSDOs) at least by some inspectors. In fact, it is being widely ignored! Many letters that go out to eligible alleged violators are sent on the old form, which makes no mention whatever of remedial training as a method of resolving the situation.

We are all familiar with the regular routine—that is, without consideration of remedial training. After the LOI comes, if the accused violator wishes he or she may have an informal conference with an associate regional counsel. At this point, unless the airman can prove that at the time of the violation he or she was in jail, the hospital, China, or some such alibi that makes it impossible for him or her to have committed the act, a sanction is imposed. The appeal from this is to an NTSB administrative law judge, and from his or her decision to the full board.

However, if the violator is offered remedial training in the original LOI, and he or she chooses to accept the offer, the entire process stops and the situation is turned over to the safety program manager (SPM) at the FSDO. His or her first step, after reviewing the investigating inspector's file on the case, is to sit down with the violator and discuss the situation. This personal, face-to-face meeting is mandatory. At this time, assuming a positive attitude on the part of the violator, a formal training agreement is signed by both the SPM and the violator. This serves several purposes. Primarily it ensures that there shall be no misunderstanding regarding the terms and conditions under which the remedial training is to be undertaken. It also protects the FAA in the event the violator fails to live up to the terms of the agreement and complete the training within the proscribed time. If this should happen, the existence of the signed agreement would prohibit the use of the "stale complaint" defense if the FAA should bring formal charges.

However, by far the most important part of the training agreement is the formal remedial training syllabus it contains. This syllabus, designed by the SPM for the specific situation, contains the duration and nature of the ground and flight instruction required of the trainee, and it is tailored to provide the maximum training benefit based on the nature of the incident which violated one or more of the regulations. A sample (model) training agreement is included as an appendix to Bulletin 90-8. Once the training agreement is signed, all that remains is for a good, competent flight instructor to administer the training, and for the instructor, trainee, and safety program manager to follow through and see that all the bases are covered.

## A case history

To see precisely how the program works, let's take an actual case and follow it through from its inception to its conclusion. It all started when a message was received at a FSDO in the southern region from an approach control facility advising that a pilot in a Cessna 172 had been "FLYING VFR WHEN IFR REQD... PILOT UNQUALIFIED FOR CONDITIONS...(airport manager)... CALLED AND ADVISED THAT N____ HAD DEPARTED AND COULD NOT RELOCATE THE AIRPORT FOR LANDING... PILOT CALLED ON FREQ____ AND ADVISED THAT HE WAS IN THE CLOUDS AND NOT IFR QUALIFIED..." ATC then vectored the aircraft to an airport where VFR conditions existed and a safe landing was made.

The file was turned over to a general aviation operations inspector at the local FSDO, and he placed a call to the pilot and left a message to contact him regarding a deviation that had occurred four days earlier. The following day the pilot returned the inspector's call, and the inspector wrote down his impressions of the telephone conversation so as to have a record for the file. The violator was asked to explain in his own words just what happened. He stated that he and his wife had planned a trip, gone to the (uncontrolled) airport in the morning, and loaded the airplane in anticipation of an afternoon departure. By the planned departure time the weather had deteriorated to the extent that the planned flight was canceled. He and his wife returned to the airport and unloaded the airplane in preparation to making the trip by automobile. Before doing so, however, the pilot decided to make one circuit of the pattern in the airplane. He stated that he knew the visibility was low, but he believed the clouds were high enough for pattern operations, so he and his wife took off. Almost as soon as they were airborne, the pilot lost ground contact in fog and lost sight of the airport. He called a nearby approach control facility and received vectors for a climb to VFR conditions on top, thence to an uncontrolled airport where a landing was made in VFR.

The inspector's written report of the telephone conversation goes on:

*I asked how many persons were on board the aircraft. He stated he and his wife. I asked what the weather was prior to his departure. He said it was a little foggy, maybe one and one-half to two miles, the clouds didn't appear to be below pattern altitude. I asked if he had received a weather briefing prior to the flight. He said he hadn't because he didn't plan on leaving the pattern.... I asked him for information concerning flight time, and flight review date for the 8020-18 Investigation of Pilot Deviation Report.... He asked what he could expect from the incident. I explained he would receive a (LOI), and it would be specific concerning procedures. He asked what the minimum penalty might be. I explained all the possible recommendations, from letter of warning to certificate action, including the possibility of remedial training.... _____ exhibited a very open and truthful attitude. He was remorseful, and has apparently learned much from the incident.*

Thereafter, the inspector completed the FAA Form 8020-18, Investigation of Pilot Deviation Report, including the statement, "Pilot attitude, and concerns have been positive and constructive and would most likely benefit from remedial training. Remedial training is being considered pending completion of investigation."

The next step was the posting (certified-return receipt requested) of the LOI, which followed the recommended model discussed above, including mention of the possibility of participation in the remedial training program and setting forth the conditions of eligibility for such participation. In this instance, the pilot met all the requirements, but he was not so informed in the LOI. This seems to be the only catch in the program. The pilot, thinking he will receive remedial training instead of sanction, may explain the situation and in doing so incriminate himself. Then if the FAA should decide not to offer remedial training, they can use the pilot's own words to convict him of the violation.

A week later the inspector received a reply from the pilot who was involved in the incident. It said, in part:

*... Although the outcome was a good one I put a lot of people to their test on that afternoon. I realize that for a five minute go around the pattern, I endangered the life of my wife and my own life. Had it not been for the excellent job of the controllers.... I realize that this matter may be under investigation and I am willing to work with you and the FAA in any way I possibly can to resolve this matter. I understand that I have an option of working with you by taking Remedial Training, if allowed, and I will receive no disciplinary action against my pilot's license. Please allow me to go to and participate in the Remedial Training Classes....*

The letter goes on to offer assurances that such an event will never happen again, and to again express his appreciation for the professionalism of the ATC personnel who rendered assistance, and again ask for remedial training in lieu of certificate action.

In following this case through from its inception to the conclusion, the next step was a letter over the signature of the safety program manager (SPM) at the Flight Standards District Office (FSDO) to the violator, advising him that he was indeed eligible for participation in the FAA remedial training program in lieu of legal enforcement action. Enclosed with this letter was a Remedial Training Syllabus and Agreement. The violator had already nominated an instructor with whom he wished to train, and the letter confirmed the approval of his nomination. The letter further states that the agreement must be signed during a personal visit between the violator and the SPM, that periodic progress reports must be made to the SPM, that the completion standards of all elements of the training syllabus must be met within 30 days of the signing of the agreement, that written documentation of this completion must be provided (in the form of the instructor's signature on the original copy of the syllabus), and finally, that "all expenses incurred for the prescribed training will be borne by you." The letter ends with the statement, "The administrative action will remain in your file for a 2-year period, after which time it will be deleted from your record."

The "Remedial Training Syllabus and Agreement" which accompanied the letter from the SPM is the real meat of the entire program. It is the syllabus that guides the instructor and the student (violator) through the training, and the training is specifically designed by the SPM for the maximum benefit based on the situation which triggered the violation in the first place. For example, the syllabus and agreement prepared for the case we have been discussing here called for 5 hours of ground instruction on those sections of the *Airman's Information Manual* dealing with meteorology and pilot/controller roles and responsibilities, several sections of the *Aviation Weather* book, those parts of the Federal Aviation Regulations dealing with instrument-rating equipment and requirements and instrument and visual flight rules. It also required that a visit be made to "an Air Route Traffic Control Center or Radar Approach Control Facility for a briefing on IFR operations within the National Airspace System." This visit must be attested to by the signature of the chief of the facility or by a duty supervisor.

Additionally, the syllabus and agreement mandated 3 hours of flight instruction, including, "a VFR cross-country flight involving flight planning and the procurement of weather information...flight solely by reference to instruments...(and)...operations to/from an airport inside a (sic) Terminal Control Area." Both the flight and ground instruction must be administered by a flight and/or ground instructor selected by the trainee and approved by the FSDO.

The syllabus also spells out the completion standards for the training as follows: "The training will have been successfully completed when, by oral testing and practical demonstration, the airman demonstrates to the supervising instructor proficiency in the above (selected) subjects and procedures in accordance with the Private Pilot Practical Test Standards." The syllabus portion of the agreement is signed and dated by the SPM.

The agreement portion, including notice of the required completion date, is signed and dated by the trainee, and finally, a completion statement is included for the signature of the instructor, who signs and dates it for the trainee to return to the SPM at the FSDO. And when this is accomplished, the trainee need only behave himself for two years, at which time the record is cleared.

In interviewing both the instructor and the trainee involved in this case, I found both to be enthusiastic about the program. A very interesting and positive point made by the trainee was the fact that he emphasized over and over again how this experience has changed his attitude toward the FAA from one of fear and apprehension to that of friendly cooperation. This particular trainee acknowledged that he needed the training and affirmed that this experience has led him to resolve to go on and undertake instruction for adding the instrument rating to his pilot certificate.

## Success or failure?

Now that we have followed a single example of just how the remedial training alternative to certificate action sanctions was accomplished, let's examine the program and see what is happening with it on the national scene.

On April 18, 1991 the Operations Branch General Aviation and Commercial Division issued a "Report Card" with a grade of A+ entitled "Remedial Training Program Evaluation Report."

Since it was determined that merely compiling the total letters of correction and relating them to specific Federal Aviation Regulations would not provide the desired detailed analysis, a questionnaire was constructed and mailed to 300 program participants. A phenomenal response to this voluntary questionnaire was experienced. One hundred fifty-one replies were received!

The 345 letters of correction received by the Operations Branch contained citations of 422 regulations as having been violated. Not surprisingly, Part 91 of the regulations, General Operating and Flight Rules, was violated most often (388 times), distantly followed by FAR Part 61, Certification: Pilots and Flight Instructors (15 citations). The remaining 19 rules violations were scattered among other FARs. The overwhelming majority of violations corrected by means of remedial training have been those dealing with the inadvertent penetration of controlled airspace of one sort or another. Altogether there were 290 such incidents out of the 422 total in the sample, 69 percent of the total violations and 75 percent of the FAR Part 91 violations. Quite obviously this type of inadvertent violation lends itself very well to the remedial training program.

Of the slightly over 50 percent return on the questionnaires sent out (151 out of 300), most (81 percent) asserted that the curriculum developed by the SPM was appropriate for their specific situation. A substantial majority of the respondents indicated a clear understanding of the curriculum objectives and understanding exactly what was expected of them in the training. Virtually all the respondents indicated that they benefited from the training. If the violation occurred other than in the airman's home district, coordination was required between the investigating inspector where the event took place and the SPM in the local office convenient to the airman, and in those cases, the respondents indicated that it was well handled. The respondents were invited to comment on the program and were guaranteed anonymity, and most of them added comments to their responses to the survey questionnaire and these comments were universally favorable.

This report-card-type evaluation of the program near the end of its first year ended with the following conclusions and recommendations:

> *The Remedial Training Program has been well-received by the aviation public and has been highly effective both in returning safer airmen to the system and in assuring continued, future compliance with the FAR. The results of the participant survey validate the program and should be construed as an indication of the need for its continuance.*

*... Because so many violations involved incursions into con-
trolled air-space, FAA needs to ensure that pilot and instructor training and certification are adequate in this area.*

In view of this overwhelming indication of early success, one must
ask why it was necessary for the Director of Flight Standards to get
out a memo on June 13, 1994 pointing out that in the first three years
there were an average of 550 remedial training cases per year, but in
year four there was a drop to 399 cases, and in the first half of year
five there were only 20 cases (and ten of them were in a single district office). This memo concludes, "All FSDO managers should reed-
ucate their inspector work force on the program's policy and
procedures...."

Obviously something is missing. Perhaps it is because established bureaucrats are prone to resist change and insist on doing things the
same old way, or perhaps because many Flight Standards inspectors
have the Gestapo mind-set that wants nothing to do with a kinder,
gentler approach to enforcement, but for whatever reason the full potential of the Remedial Training Program is not being approached.
Airmen should know that they have an absolute right to be informed
regarding the program. Understand, they have no right to participate
in the program, but they do have the right to be told about it in
the original LOI, and even this isn't being done in many instances.
If more airmen were aware of this alternative to punitive action and
more inspectors were reminded of their duty to try to use it, we
would no doubt see a resurgence in its use. If the FAA wants to
change its image from one of adversarial to cooperative, if they really
want their customers (the airmen) to believe it when they hear, "Hi,
I'm from the FAA and I'm here to help you!," then more use of the remedial training alternative to help ensure compliance is one means of
achieving this end. If this program becomes more widely used, re-
placing certificate action and/or civil penalty sanctions, everyone involved would not only benefit, but also would be happier.

I have already cited a few horror stories regarding what can happen
to an airman (man or woman, pilot or mechanic) at the hands of the
FAA. It is all done in the name of safety, but many airmen have
found that some little mistake, having absolutely nothing to do with
safety, has been caught up in the FAA enforcement machinery, and
airmen have suffered grievously because of it. Is it any wonder, then,
that the aviation community has come to look upon the FAA as the
enemy?

# 14

# What now?

Why do you fly? If you are reading this it is because you are a pilot. What motivated you to become a pilot? Are you a professional pilot, professional wannabe? Do you fly for business or just for pleasure? Whatever got you into aviation, whatever level you occupy, you are no doubt looking ahead.

What do you want to do next? We all need a goal toward which we can work as we strive to improve our technique. For presolo students, it is the day they experience their first solo flight, then it's the first solo cross-country flight. For VFR private pilots, it may be their first truly long trip, or perhaps the instrument rating.

## I The flight review

Twenty-five years ago the FAA selected a general aviation operations inspector in each of two district offices (Detroit and Grand Rapids districts in the Great Lakes Region), designated them accident prevention specialists, and charged them with the responsibility of establishing a program on a test basis. Each of these two inspectors selected several flight instructors in his district to work with him as volunteer accident prevention counselors, and they set to work developing a workable safety program for pilots. (One of these two inspectors recently retired as the safety program manager, formerly accident prevention program manager, and before that the title was accident prevention specialist, in the Minneapolis District Office.) At the end of the test year, the concept had proven worthwhile, and the accident prevention program was born. It went national, and one inspector in every district office was chosen to act as accident prevention specialist.

One of the features of the original program was the "safety pin," a silver (fake silver—I think it was pewter) lapel pin in the shape of a safety pin with the "spirit of St. Louis" in the center. This pin was awarded to any pilot who took a "voluntary proficiency checkride" with one of the accident prevention counselors or the accident prevention specialist, as they were called then. After a few successful years of this, the voluntary proficiency check became mandatory, and the biennial flight review was born. (By the way, the "Wings" program is a sort of extension of the voluntary proficiency check as well. More about this later.)

The thinking of the powers that be at the very top in the Flight Standards section of the FAA was and is that since professional pilots (particularly air-carrier and military pilots) are required to get retraining and proficiency checks, so why shouldn't the general-aviation pilot as well? The casual or infrequent flyers no doubt need to have their skills (and knowledge) reviewed more than does the professional who is flying every day and is likely to be staying abreast of what's going on in aviation.

From its very inception, the conduct of the BFR has been left entirely to the discretion of the administering instructor. This, of course, has resulted in the review consisting of anything from a perfunctory "once around the patch" to several dual instruction sessions involving several hours of square business flying. The FAA has carefully refrained from specifying specific maneuvers or procedures in the conduct of the review. In other words, the FAA has imposed no requirements on the instructor, but it has offered suggestions in an effort to standardize the procedures as to just how the review might be conducted. In what seemed like no time at all, however, as soon as the review became regulatory virtually every commercial publishing house in the business was out with a checklist of material to be covered on a BFR. This included Jeppesen-Sanderson, which, putting out one in their own name, prepared the *Cessna Pilot Center Flight Review Manual* (and filmstrip), among others. Even the AOPA Air Safety Foundation publishes a "Flight Review Checklist" which includes a comprehensive review of all the private maneuvers and procedures. These guides for instructors conducting a review run the gamut all the way from a simple 20-item checklist to the Cessna Pilot Center "Flight Review Manual," which is a 70-page booklet.

Of course, the definitive guide for the flight review is the FAA Advisory Circular, AC 61.98A. Although this is a description of recommended procedures for the conduct of a flight review in text form, it is, in effect, a textual checklist which the instructor may use as a guide in the conduct of the review. And one of the many FAA accident prevention publications is a checklist to guide the instructor administering a review. The review itself is not meant to be a pass-fail situation, but without a "satisfactory" endorsement from the instructor (reviewer), the applicant (reviewee) may not legally operate as pilot in command of an aircraft (even all by himself or herself, alone, solo). Therefore, despite protestations to the contrary, every pilot must "pass" the review. Originally the expiration of the time for a review was the anniversary date two years after the last one, but now, like everything else in the FAA, the review expires at the end of the twenty-fourth month since the last one.

When the flight review first became mandatory, I heard an FAA aviation safety inspector addressing a large group of flight instructors say, "You all now have an additional opportunity to make money. Remember, if pilots who are overdue for a BFR taxi across the field to meet you, it's okay, but if they have to fly to your airport, they're in violation and you got them!"

An instructor in attendance stood up and said, "Maybe you got them, but I don't. If they're conscientious enough to come for a review, I'm certainly not going to report them!"

Since the purpose of the review is to ensure that all pilots maintain the level of proficiency they had reached at the time they acquired the certificate in the first place, the ideal review involves the use of our old friend the Practical Test Standards as the guide for the conduct of the review. And the FAA is now pushing this as the guide for an instructor giving a review. If the pilot fails to meet the standard set forth in the appropriate PTS, the instructor should not issue him or her a "sat" on the review, but rather merely sign his or her log as having received dual instruction for that session, and require another period (or more) until the pilot once again achieves the standard of the PTS prior to signing him or her off as having satisfactorily passed a flight review. Let the PTS be the guide. After all, that's what applied in the first place. Or did it? Perhaps the pilot acquired his or her certificate under an old, different standard, in which case maybe the pilot should be "grandfathered" in under that standard. You can't very well ask the pilots to be better than they were when they got the certificate.

Personally, when I administer a flight review I hold the pilot to the standard of the appropriate PTS, but I select only those procedures and maneuvers that in my judgment will be most beneficial for the pilot. I always include flight at minimum controllable airspeed while maneuvering the airplane and maintaining a precise altitude, a stall series (with emphasis on stall recognition as opposed to a mere demonstration of stalling the aircraft and recovering), and several landings, including crosswind, short and soft field, no-flap landings, and slips. We usually also do one or more maximum-performance takeoffs. Additionally, I give an oral, including as a minimum airspace regulations, fundamental aerodynamics, and weight and balance.

I'm an absolute nut on the subject of weight and balance. I spend a great deal of time providing expert testimony in legal actions involving airplane accidents, and in the vast majority of aircraft mishaps, weight and balance is a factor. It may not be the determining factor, but it is a factor, and often it is the cause of the accident; and when it is, it's almost invariably fatal. If the pilot seeking a review holds a commercial certificate, I will have him or her do a few of the commercial maneuvers (chandelles, lazy eights, and steep power turns) because the skill level in terms of maneuvering the airplane smoothly and precisely is what basically separates the commercial pilot from the private.

By the way, one can tell a great deal about the skill of a pilot by just how well he or she does steep power turns of 360 or 720 degrees. The depth to which I go in the administration of a flight review depends on just how well I know the individual (as an aviator, not as a person). If he or she has been flying regularly and to my knowledge been staying abreast of what's been happening in aviation, I don't give as comprehensive a review as I do for the infrequent flyer (or the individual with whose skills I am not familiar). And although the review is supposed to be an evaluation rather than a lesson, I certainly hope that each time I give a review, the pilot enjoys a learning experience.

Even so, probably the most violated of all the Federal Aviation Regulations is Part 61.56 (formerly 61.57), which sets forth the requirement of the flight review. It seems to be easier for the pilot to forget this than the need for a medical every 6, 12, or 24 months, as the case may be. Then there is that large group of pilots who have the infamous antiauthority attitude and maintain that they have nothing to prove and refuse to expose themselves to an instructor for a review of their skill and ability. After all, they did it at the time they acquired the certificate, so why should they have to do it again? The irony of this is that these are the pilots who need it the most! The serious, safety-minded pilots who are very current in their flying and who conscientiously present themselves to an instructor for a review every 24 months (or more often) are the ones who will benefit from the review the least. They probably don't really need it; they certainly don't need it to the extent that the pilots with chips on their shoulders who refuse to expose themselves to the scrutiny of an instructor need it.

The most recent changes in the evolution of the flight review regulations include:

- The on-again, off-again, on-again annual flight review for recreational or private pilots who have less than 400 hours total time flight experience, or, in the case of private pilots, do not have an instrument rating on their pilot certificates.

- Requiring a minimum of one hour of ground and one hour of flight time on the review (annual or biennial).

These refinements in the requirements of the flight review have been proposed, abandoned, and reproposed, and now, for the first time, the discretion of the administering flight instructor has been restricted by the imposition of a minimum time requirement. But this is under review by the regulators again.

Another proposal for change in the review requirements, and this one, too, has fallen by the wayside (fortunately) called for the review to be accomplished in the most complex aircraft the pilot is qualified to fly by virtue of the certificate he or she holds. Then there was the proposal that every pilot be reviewed in every category and class of aircraft for which he or she is certificated. We are, however, still in the position of permitting pilots who hold an ATP Certificate (with several type ratings) for Airplanes Single-Engine Land and Rotorcraft-Helicopter, Commercial Privileges for Airplanes Multiengine Land and Single-Engine Sea, and Private Privileges for Gliders and Lighter-Than-Air Hot-Air Balloon with Airborne Heater to take the flight review in a glider or a balloon, and it covers them in all the aircraft in which they're certificated.

There are several other ways of satisfactorily accomplishing the required flight review, in addition to the pilots demonstrating to a certified flight instructor that they can still meet the standard to which they flew when receiving their certificates. The acquisition of a new grade of certificate or rating on a certificate counts as a flight review, but the applicant must fly. (For example, CFIs who renew their instructor certificates without actually flying with an examiner or inspector still require a flight review, while instructors who renew as a result of a flight test are covered as having been reviewed.) In fact, any required checkride counts as a flight review, and this is true whether or not it is a pass-fail situation or merely a standardization ride (not a pass-fail deal).

Another way (fairly new) to get credit for having completed a satisfactory review is to participate in the WINGS program. This program, now entering its sixteenth year, has enjoyed nothing less than phenomenal success. Any certificated pilots (above student grade) who attend an FAA Pilot Education (Safety) program and within 365 days (formerly 60 days) of doing so receive 3 hours of dual instruction—including 1 hour of takeoffs and landings, 1 hour of air work (maneuvers, stalls, slow flight, etc.), and 1 hour of instrument instruction—are awarded an attractive lapel pin (in the shape of a pair of wings, hence the name of the program) and a certificate attesting to their interest in safe flying. Once each year, the pilots may repeat the process (attend the safety seminar and get the 3 hours of instruction), and upgrade their WINGS from the present stage to the next stage up the ladder. The pin itself is a derivative of the military wings, from Pilot Wings (with a shield) through Senior Pilot (with a star above the shield) to Command Pilot (with a wreath around the star above the shield), and it makes quite an attractive lapel pin. And of course, each stage (or, more properly, phase) of the WINGS program offers successively fancier wings pins for the lapel. And now, pilots can get credit for having attended a safety meeting by sitting in on the Aviation Forum safety meeting conducted by "Buz" Massengale by computer on America Online (Wednesday nights at 10:00 eastern).

Human beings have a need to know that what they do makes a difference, that there is some means of measuring the results of their efforts, and for several years I felt a great sympathy for the accident prevention specialists who could put forth a monumental amount of work and never know that they prevented a single accident. In our local FAA District Office we have had nine accident prevention specialists (or APS-APPM-SPM) over the years, plus several who acted as APS for one month only. (While we were between APS, the office manager assigned each general-aviation operations inspector to fill the position for one month.) While I was sitting in the office of one of these guys, he kept reaching over and making a tally mark on his office wall with a piece of chalk. On the fifth he'd make the usual hash mark across the first four. I asked him what the hell he was doing, and he said, "Look out the window there. Another airplane just landed, and I prevented another accident!"

The first actual measurement of success in this area came with the advent of the WINGS program. It is my understanding that after the first full year of this program, some 20,000 pilots had earned the first-stage wings, and so far as we know, not a single one of 'em had been involved in a reportable accident, incident, or violation. This is an extremely impressive record, and has made me a vigorous supporter of the program.

By the way, one thing that does not qualify as a flight review is an instrument competency check, which, although required of the pilots who have permitted their instrument currency to lapse, is only a check of their ability to fly on the gages and operate in the IFR System and is not a check of their skill level as related to the GRADE of certificates they hold. Therefore, an ICC cannot be offered in place of a BFR. They are two entirely different kinds of checks. Of course, the two may be combined in a single transaction, and they frequently are.

In considering a flight review, perhaps pilots, and particularly flight instructors might ask, "Just how good should pilots be?" Once again, our friendly PTS comes into play. The answer, of course, is, "At least good enough to meet the minimum standard as set forth in the Practical Test Standards for their grade of certificate, a standard they met at the time they were originally tested"

## II  The rating collectors

Many pilots seek advanced ratings "because they're there" and for no other reason. These are the guys (and gals) we used to call rating collectors. They are the ones who acquire an instrument rating never intending to use it, a commercial certificate never intending to fly for hire, and an instructor certificate never intending to teach. They even go for an Airline Transport Certificate as soon as they meet the minimum experience requirement. However, the true "rating collectors" are the ones who are just looking for more words on their pilot certificates. Those in the first group may very well be seeking advanced ratings as a means of improving their skills.

I used to belong to a glider club and one of the members was a pilot for a major air carrier who was also a glider instructor who donated his talent to instructing student members. As he traveled about
the country in his capacity as an air-carrier pilot, he posted a notice
in pilot lounges throughout the system, offering a special glider transition course. He got quite a few takers among the ATP air-carrier
pilots who read his notice, and when these guys would complete
their brief transition glider course (from power pilot to glider pilot)
I would get them for their certification practical test. In some cases I
would have to write a three-page and on one occasion even a four-
page pilot certificate to get all the ratings and type ratings in. In several cases this included multiengine seaplane ratings, hot-air-balloon
ratings, and helicopter and numerous airplane-type ratings. That's
what I call collecting ratings for no valid purpose other than that
they are there, that they add additional words to the pilot certificate.
These are the true rating collectors. Once they achieve the multiengine seaplane rating, they will never again fly a many-motor seaplane or amphibian.

We should all set goals for ourselves. We should all be constantly
striving to improve our skills and abilities. And if advanced certificates and ratings are the goals that pilots set for themselves, that's
fine, but pilots who think the ATP entitles them to some sort of
bragging rights are sadly mistaken. If by the time pilots have
achieved 1,500 hours of experience — including the cross-country,
night, and instrument experience to fulfill the requirement for the
ATP — they have not also acquired the knowledge and skill to pass
the flight test and acquire the certificate, then they have been wasting all those hours tooling around the sky without making an effort
to do so precisely and smoothly, without challenging themselves to
improve their skills. And this is indeed a waste. If we're not striving
to get better, we get worse. We become sloppy and complacent,
and there's no place in aviation for sloppiness or complacency.

## III  Written tests

For those of you who do intend to seek advanced ratings, if there
is a written examination required, pay close attention while your
Old Dad gives you his handy-dandy method of cheating on the
writtens. No, I don't mean actually cheating, but the result is
the same. If you follow the procedure I'm about to lay out for you,
you are guaranteed to pass with a high score.

The first step is to know the material. Do not study the test; study the material. It is my firm belief that if you know the material the test will take care of itself. Then, when you sit down to take the test, carefully review all the supplementary material (data for the hypothetical airplane you will be using, etc.). Next go through the test, rapidly answering all the questions that you know cold. Don't waste your time sitting there scratching your head. The knowledge won't just come to you. Don't do any problem solving. In other words, the first time, through skip all questions that involve problem solving or about which you have any doubt. When you finish going clear through the test this first time, you will have answered some 60-odd percent of the questions, and answered them correctly because they were all stuff you know without doubt.

Now go clear through the test again, solving all the problems that require a flight computer or plotter. This time you will also discover that the answers to a few of the questions you skipped the first time through showed up in later questions, so you can answer them as well. Again skip all the questions to which you just don't know the answers. By the time you complete the second trip through the exam, your score will be in the high 80s. Finally, go through the test one more time, this time applying logic to get the answer. Of the three choices, one is obviously wrong. Eliminate that one. As between the two remaining choices, pick the one that seems best, and if you can't tell which it is, flip a coin! If you follow this procedure you are sure to score in the mid to high 90s.

# IV  Flight instructors and the VA

The ranks of flight instructors are filled with people who have earned the certificate with no intention of ever using it. During the days when the military services were releasing members of the armed forces by the thousands and the Veterans Administration was paying for flight training, literally thousands of veterans took advantage of the opportunity offered by their "rich uncle" to get all the flight training for which they were eligible, and this included the flight instructor certificate. They used their VA eligibility to go joy riding in the sky, never intending to teach. They got multiengine ratings and never again flew a twin. They got seaplane ratings and never again flew on floats.

After the Second World War, Uncle Sam paid for all this and there was a great spurt in flight training. So-called "VA flight schools" sprang up all over the country. However, it wasn't too long before the government wised up and declared that it would only pay for vocational flight training. Veterans had to claim that they intended to earn their living as pilots before the VA would pay for their training, and to ensure that the veterans were serious, Uncle Sam would only pick up 90 percent of the cost, and that only after the applicants had acquired the private pilot certificate on their own.

So what happened was those veterans who wanted to claim their benefits for flight training were forced to lie. We had doctors and lawyers with established practices stating that they intended to throw away all their previous training and their established professional situation and become professional pilots. Although they were forced to say they were going into aviation as a vocation, it was, is, and always has been an avocation for them. These people collected an awful lot of ratings that were never used. They became rating collectors by inadvertence. This is different from those people who acquire a whole bunch of certificates and ratings just so they can claim they have them. The only thing wrong with this is that it gives an unrealistic picture of how many pilots of each level there are, and I'm not even sure that this is necessarily bad, except that it does distort the statistics. I wonder just how many ATP-certificated pilots there are out there who have never been in the position of flying in a situation where the ATP is required?

It is my belief that the very highest pinnacle to which aviators might aspire is to be a truly great primary flight instructor. There isn't an astronaut who stepped on the moon, there isn't a high-time senior captain herding one of those flying condominiums across the big puddle who didn't get their first flight lesson from a primary instructor. We all started with a primary instructor, and it was from him or her that we acquired the habits that have carried forward through our entire aviation careers. It is unfortunate, however, that the ranks of flight instructors are filled with ambitious young aviators who are using their CFI certificates as stepping stones in their own careers, while they accumulate the experience (read hours in a logbook) to make themselves attractive to the corporate or air-carrier employer, and they are doing it at the expense of their students! They don't care a whit about the welfare of their students. They consider instruction "paying their dues," an unwanted step along the way to their real goal. This is a built-in flaw in the aviation education system.

Then there's that group of part-time instructors (by no means a majority of part timers) who like to fly, but the only way they can afford to do so is if somebody else is paying the freight. Flight training by nature must be a hands-on proposition, but instructors of this ilk rarely give their students the opportunity to fly the airplane. They persist in demonstrating each maneuver and procedure over and over again, enjoying themselves and having fun flying the airplane while the student helplessly sits by and watches and learns nothing but how good a pilot the instructor is (or perhaps is not).

Please do not misunderstand me. This is not to say that there are not some very good, dedicated instructors out there doing it only until they can get a "real" job in aviation. And there are lots of fine part timers who love to teach and are very good at it, but can only afford to do it on a part-time basis. To them, the money they earn instructing is of very little importance. They have other occupations or professions which provide their livelihood and are instructing because they love aviation and enjoy sharing it with others.

The bottom line is: if you find good instructors who can teach and fly, latch on to them, treasure them, and, for heaven's sake, pay them what they're worth!

And on the subject of paying your flight instructor, I have observed an interesting but totally incomprehensible phenomenon. Many owner pilots with the world's supply of money will spend thousands of dollars on some useless toy, but stint on instruction as they shop for the cheapest they can find. What is it with these people? Have they no respect for the value of good teaching, or are they merely contemptuous of other human beings? This kind of activity is similar to the corporation that will spend countless thousands on cosmetics (unnecessary paint and interior work) while neglecting really important stuff that their airplane could really use, such as an autopilot, radar, GPS, or other avionic equipment that would render the pilot's job easier and the corporate airplane more useful. This is, of course, a result of the fact that the money people in the company are ignorant regarding aviation and are interested only in catering to the comfort of the top executives who ride in the company planes. In this case, it is up to the staff in the flight department to educate the bean counters. However, in the case of the flight instructors, it is up to them to insist on being paid what they are really worth. In what other profession is the investment in becoming qualified so high and the return so low?

# V The challenge (what to do next)

In Chapter 10 we discussed flying different kinds of equipment. If you undertook flight training to meet a challenge and have reached that first goal, why not go for a glider or seaplane rating, or undertake aerobatic instruction as your next goal? There are a myriad of opportunities out there just waiting for you. If you have a competitive nature, you can get serious about competitive aerobatics, or you can become a proficient sailplane pilot and enter glider contests. All it takes are time and money: time to practice until you're really good, and then practice some more, and money for equipment and travel to the tournaments.

## Good work

Most pilots who have a love for aviation are anxious to share their joy with anyone who is willing to put up with their enthusiasm. And there are a great many ways in which this can be accomplished. There are numerous programs in which we can introduce others to the wonders of aviation, one of which is the First Flight Program.

## First flight

The General Aviation Historical Association has taken the Young Eagles program of the Experimental Aircraft Association a step further. On June 19th, 1995 the first annual First Flight program got underway. Sponsored entirely by the GAHA (a nonprofit, tax-exempt corporation chartered by the Commonwealth of Pennsylvania) and using donated airplanes, six young people with a demonstrated interest in aviation were given approximately 10 hours of dual instruction in tailwheel airplanes. The entire project is the brainchild of John Shreve, who donated countless hours of time and dollars of money to get it off the ground (literally).

It was hoped that they would all get to solo, but because of the weather only three of the six soloed during the time the program ran. I was asked by John C. Shreve, who dreamed up the idea, to be Director of this first First Flight program. Mr. Shreve had arranged for the use of two Aeronca Champs, one Piper Cub and one Aeronca Chief, all offered by their owners without charge. Mr. Shreve owns a resort in the Pennsylvania mountains, with a restaurant, swimming pool, other recreational facilities, and a classroom, and most important—a 2,600-foot sod strip. (See Fig. 14-1.)

**14-1** *Instruction staff, left to right, Gary, Harold, Howard, and Pete.*

The GAHA provided meals, lodging, and fuel for the airplanes, and I recruited three experienced taildragger instructors to teach the flying—two students per instructor and per aircraft. We held the Chief in reserve, preferring to use the two Champs and the Cub because of their tandem seating. I taught the ground school and gave stage checks and graduation rides to the students. The students were all Civil Air Patrol Cadets, and their selection was left up to the Pennsylvania Wing of the CAP.

The airport, called Shreveport North (62PA) is on three sectional aeronautical charts—Detroit, New York, and Washington—and the strip is oriented Southwest to Northeast (Rwys 6 and 24). It has nice, long, clear approaches at both ends. There are two public-use airports within 3 miles of Shreveport North, one so close that its pattern intersects that of Shreveport North. However, by using a right-hand pattern at Shreveport North, the two are kept separate. Using runway six for takeoff, by the time pilots are 500 feet above the ground, they can glide straight ahead and land on a runway oriented exactly the same as the one from which they took off. The two airports are separated by only a single field (also a suitable landing area in case of emergency).

The six students (five boys and one girl) ranged in age from 16 through 19. They were required to have a third-class medical certificate, a student pilot certificate, and a copy of the FARs. The Commonwealth of Pennsylvania, through its Aeronautics Commission, donated state aeronautical charts and Pennsylvania Airport Directories to the program. Although none had previous flight experience, all the kids were, of course, serious aviation students and this made the job of the instructors a lot easier than it might otherwise have been. Although at their age it is a bit early to be sure, I believe that most, if not all, of them will make careers in aviation.

The students were housed in a bunkhouse, a barrackslike building with individual cubicles which afforded a degree of privacy, and we were all fed at the restaurant. Three of the six cadets, in alternating shifts, worked in the kitchen at each meal under the supervision of a master chef. They also had the duty of waiting tables for the rest of us and clean-up duty. The meals were superb. (Figs. 14-2 and 14-3.)

After a hearty breakfast (served promptly at 7:00 a.m.), we had a ground session in the classroom each day. Then the day's flying began. We broke for lunch at noon, and this was the big meal of the day. After another ground session at 1:00 o'clock, the afternoon was spent flying. Supper at 6:00 was hot sandwiches, cold beverages, and ice cream served from a chuck wagon right out on the flight line. Then more flying as long as daylight and weather held out.

**14-2** *Ground school.*

**14-3** *More ground school.*

The curriculum I designed for this program was based on the one we used when my flight school had a contract to train Air Force ROTC Cadets, excluding, of course, communications with a control tower and cross-country work.

Fortunately, the first two days we had good enough weather to get in all the required high work (stall series, etc.) because the last half of the week we stood down more than we flew. With low ceilings and limited visibility, we sat around the flight line and every time it lifted enough to work in the pattern, we flew. We used a 600-foot pattern and full-stop taxi-back landings. (Since a substantial portion of taildragger training is learning how to manipulate the airplane around on the ground, touch-and-go landings are not really suitable.) By Thursday all the students had passed their required presolo writtens, and by Friday they were about ready to solo.

Saturday we sat around on the flight line in the rain and drizzle, and every time there was a pause in the precipitation we'd get a little flying in. By mid afternoon three of the six had soloed, and it became obvious that there would be no more flying for the next few days, so the program ended Saturday afternoon. It is unfortunate that they didn't all get to solo, but the objective of the program was definitely met. The two cadets who had been flying the J3 Cub were denied the opportunity to solo because their airplane developed an ignition problem on Friday evening, and it was felt that at that stage it would have been too much of a transition to put them in one of the Champs or even worse, in the Chief. They had been training in the rear seat of the Cub and would have had to learn from the front seat in the Champ, and the Chief is configured as a side-by-side as opposed to the other two, which are tandem.

It is hoped that this First Flight program will continue for many years to come. All of us who participated in teaching these fine young people agree that we all derived a keen sense of satisfaction and were well rewarded for our efforts. Every one of us is looking forward to the future, when we hope to be able to repeat the program. It is also hoped that this activity can be expanded to other areas of the country.

## The Civil Air Patrol

Doubtless the most well known of people who share their love of aviation is the CAP. In 1941 the Civil Air Patrol (CAP) was established as an agency in the Office of Civilian Defense. It started out as an organized group of volunteer civilian pilots and others with an interest in aviation and civil defense. The pilots, with an observer along, flew their own planes in courier service, antisubmarine patrol, and search-and-rescue (SARCAP) missions. In 1948, after the Second World War, the Civil Air Patrol was attached to the newly formed United States Air Force as its civilian auxiliary.

Today, the CAP, headquartered at Maxwell Air Force Base in Alabama, is still flying search-and-rescue missions, but it has grown to almost 70,000 volunteer members, nearly half of whom are young cadets. It is organized by states, within which are Wings, Groups, and Squadrons, some of which own airplanes, but most of the 10,000 airplanes available to the CAP are member-owned. The Civil Air Patrol maintains a nationwide radio network, and its members fly thousands of hours annually searching for downed or missing aircraft and participating in rescue missions.

It is the official Auxiliary of the United States Air Force, and its three main missions are:

1. Emergency services (missing aircraft, etc. including disaster relief)
2. The Civil Air Patrol Cadet Program
3. Education of America at large in aerospace knowledge

If you are a pilot—whether or not you own an airplane—and you have the time to donate to this good work, you would be well advised to consider joining the Civil Air Patrol. No doubt there is a squadron nearby, wherever you are, and if you are a parent, the Civil Air Patrol is a great means of keeping kids occupied and otherwise out of trouble. And they learn lots of useful stuff.

## Volunteer medical flights

There are several organizations through which people who own and fly their own airplanes can volunteer to carry medical patients and/or their families to distant destinations for highly specialized treatment. This is completely different from the so-called medivac programs that serve many hospitals and other medical facilities. This kind of operation is based on a contract with a local helicopter charter outfit that rushes emergency medical cases (usually accident victims, but other sorts of medical emergencies as well) to the facility. What we're talking about here are the volunteers who take patients relatively long distances to receive treatment or surgery. And they sometimes carry the parents of child patients to the facility where their children are being treated.

There are several such organizations through which owner-pilots can volunteer their aircraft and their time. One such is Angle Flight of Virginia, operating under the banner of Mercy Medical Airlift. This organization provides charitable medical air transport to people in need. Free air transportation is made available to patients with special medical requirements and their families. The owner-pilots volunteer their time and the costs of operating their own airplanes. The normal distances flown are up to 500 miles or so, but through Mercy Medical Airlift, arrangements can be made to connect with adjacent organizations to handle multileg flights.

## Air racing

After flying for well over 50 years, I recently discovered a whole facet of aviation of which I had not been aware. I am referring to handicap racing. There is a regular air race season, and a substantial number of people travel around the country entering a series of these races. If you are looking for a challenge—an opportunity to test yourself against an empirical standard—this is it.

As a boy growing up in suburban Cleveland during the period that progress in aviation was measured by air racing, I attended the

National Air Races throughout the 1930s and early 1940s, and I was privileged to meet all the aviation greats of that era (Jimmy Doolittle, Rosco Turner, Amelia Earhart, etc). The Bendix Trophy Race was run from California, and the first plane to land at Cleveland Municipal, now Cleveland Hopkins Airport, won. The big race, the Thompson Trophy Race, was flown around a closed course, right in front of the grandstands, around a set of pylons in plain sight of the crowd, and again, the fastest plane won. And this was my concept of air racing right up until May of 1996, at which time I experienced an entirely different kind of racing — the handicap race.

I was fortunate enough to be asked to fly with a former student at my flight school and friend who has been racing his V35A Bonanza for several years. He was entered in the Great Southern Air Race and needed another pilot to fly with him as copilot. Henry is quite an experienced and successful racer and accompanying him was a pleasure and a privilege which afforded me an opportunity to learn a great deal more about handicap racing than I otherwise might have.

We flew Henry's Bonanza from Michigan to Vero Beach, Florida, on Sunday, May 19, 1996. The airplane is loaded with virtually everything you can put in such a machine. In addition to color radar, it has a KLN 90 GPS (essential for the race), a Flight Director, an Argus 5000 moving map, HSI, 3-axis autopilot with altitude preselect, fuel totalizer, two ADFs, a DME, Sky Phone, and lots more (including a HF radio which Henry used when he flew to Europe and South America). After landing at VRB on Sunday, we registered and had all the paperwork checked, both the airplane's and our own. (Figs. 14-4 and 14-5.)

On Monday the aircraft was inspected and everything removed to make it as light as possible for the handicap run, which was accomplished with only the pilot, the official handicap pilot, and full fuel aboard. The handicap run is over a course with a timer on the ground at each end. It is flown at full takeoff power, and a handicap speed for each racer is established. Since the race is run against the speed index thus established, the slower the better. After the handicap run, the airplanes are impounded until the race starts, thus rendering it impossible for the racers to make any changes to their equipment which would give them an advantage. Each airplane has its assigned race number prominently displayed so the timers and spotters can issue proper credit at the flybys and start and finish lines.

**14-4** *Howard and Henry.*

Once the race commences, all radio calls are made identifying the airplanes by their race numbers, "Racer two-two, five miles out," etc.

Altogether there were some 30-odd volunteer timers and spotters working throughout the race. The 1996 race was the twelfth annual running of this particular event. There were a total of 41 airplanes registered for the race this year, and 39 actually flew. They included Cherokees, Grumman Americans, Cessna 172s and 182s, Mooneys, Beechcrafts (our 35 and a couple of 36s), and assorted twins, including a Beech Baron, a Cessna 310, and an Aerostar. No turbocharged or experimental airplanes are allowed. Because it is a handicap race, any airplane has as good a chance as any other to win. Past winners have included a Piper Dakota, a Cessna 172, and even a Cessna 152. Since the president of the Florida Race Pilots Association (the organizing body), Cheryl Finke, was racing this year, an individual was hired to supervise the entire affair. John Walker did an outstanding job of seeing that the whole thing ran smoothly.

Monday afternoon there was a mandatory pilots' briefing at the headquarters hotel, which included a safety talk by a U.S. Coast Guard officer, after which ICAO International Flight Plans were filed. That evening there was a welcome reception at the hotel, with hors d'oeuvres, etc. (There was some kind of party every day.)

**14-5** *Racer number 69—Howard and Henry.*

The race was scheduled to start on Tuesday at 9 A.M., and we were supposed to fly to Marco Island, Florida, North Palm Beach County, Florida, Walker's Cay, Bahamas, Freeport, Bahamas, and end up at Bimini, Bahamas, for a mandatory overnight stay. However, the weather over peninsular Florida was rotten, so we ended up taking off about noon for Freeport (MYGF), where we cleared Bahamian customs and fueled up. Therefore, the first actual race leg was from Freeport to Bimini.

The fastest airplanes take off first so as to prevent a jam-up of over-taking airplanes. After takeoff, each airplane flies a wide, modified pattern and dives down across the start line parallel to the runway at either 300 or 400 feet AGL, resulting in maximum speed as they start on course. The start line is usually about midfield, and the racers must fly the entire length of the runway prior to turning on course. (There is a spotter sitting at the end of the runway to make sure no-body cuts the corner and turns on course until he/she has passed the end of the runway.)

As soon as the racers pass the departure end of the runway, they turn on course for the next flyby, and they head out at full takeoff power. We had the throttle wide open and the rpm sitting right on the red-line. The racers stay low, 50 feet or less over the water, in order to take advantage of ground effect and to avoid wasting time climbing to altitude. (See Figs. 14-6 and 14-7.) On the first race leg, we encountered several light-to-moderate rain showers, and the radar painted a couple of good-sized cells off to the side. The rain wiped out our stuck-on race number and that of several other racers.

The finish line for that first day of racing was an imaginary line across the runway at Bimini (MYBS), and after crossing the finish line where a timer clocked us in, the racers circled around and landed for an overnight stay in Bimini. Two of the flybys (turn points) were lighthouses, and volunteer timers or spotters were on site at each to make sure the racers didn't cut a corner, but flew around the lighthouse.

Wednesday, the second day of racing, we flew from MYBS to Marsh Harbor, where we overnighted. Thursday was supposed to be an off day at Marsh Harbor, but because of being behind schedule it was a race day—two legs, and another night at March Harbor, with the usual reception and party. The accommodations at each of our stops were superb. First class all the way.

Friday was the last day of racing and we flew from Marsh Harbor by the Hopetown Lighthouse, Elbow Cay, Bahamas, Governor's Harbor (Eluthera), Bahamas, Chub Cay, Berry Islands, Bahamas, finishing as we flew by the Lucayan Lighthouse, Freeport (Grand Bahamas).

**14-6** *We were 40 feet above the ocean.*

**14-7** *Number 3 was 20 feet below us! (How's that for flying on the deck?)*

We overnighted in Freeport, and Saturday was an off day while the scores were being computed. By then I really needed the day off to rest up from the tension of racing full-bore so close to the water. The top several winners were literally within seconds of each other. Henry and I finished high in the middle of the group, and we would have done much better had we not experienced one very bad leg. Even so, we won a portion of the $25,000 prize money for firsts and seconds on a couple of legs. Saturday evening the awards banquet was held in the hotel at Freeport, and there were prizes for everything imaginable (I even won a prize for either being the oldest entry or the one with the most flight experience).

This was the twelfth annual Great Southern Air Race, and it was won by William Waldron of New Jersey flying a 1965 Comanche. It was his first race! And he took home $5,000 for his first-place finish. The father-and-son team that took second-place prize money of $3,000 was a couple of fellows that I had flight checked for their seaplane ratings several years ago. Marco and Lucky Pierobon of Michigan were in the race for the fourth consecutive year, and this time they flew a Cessna 180.

On Sunday we all started home, most of us stopping at Fort Pierce, Florida, to clear U.S. customs back into the states. Altogether, it most certainly was an interesting experience for me to participate in this race, but I'm not sure I would want to do it again, at least not right away. Ask me next year.

While it's not cheap to race, you don't have to be rich to do it, either. Entries and other required fees (you have to belong to the club) added up to just under $300 per team, and trip expenses for us amounted to about $3,000. So with a top prize of $5,000 at stake (total prize money for the race was $25,000), high finishers would win back their expenses and then some.

The Great Southern Air Race is one of the most popular, and it certainly offers the largest amounts of prize money, but there are lots of others. Some other races establish handicap indices as described above, while others use the manufacturer's performance numbers for their handicap. Many of the races around the country are sponsored by the Ninety-nines, the international organization of women pilots.

## The really long trip

Another kind of challenge, and a real adventure, is to plan and execute a trip to Europe, Alaska (if you live in the "lower 48"), or Central or South America. Books have been written on flying to and in each of these places, so I need not describe in detail the equipment and procedures required. Suffice it to say it can be a real adventure.

Three times I have come very close to flying across the Atlantic, and someday I hope to actually do so. At one point I was scheduled to deliver an airplane to Belgium, and I had gone so far as to make all kinds of preparations when the buyer changed his mind and decided to fly it home himself.

There are a whole bunch of good books written describing the procedure for flying to Alaska, overseas, in and around the Bahamas, and Hawaii, so I need not go into any great detail here. I have flown all around the Hawaiian Islands and the Bahamas in single-engine airplanes, and I can tell you it is quite a thrill to be low over open water out of sight of land when you sight a speck on the horizon and it gradually grows into the island you have been aiming for.

# 15

# Other considerations

## I Prejudice in aviation

The important thing to remember is that the airplane doesn't know or care who is flying it. One of the things I like about aviation is the fine bunch of people who do it. On the whole, pilots universally respect the skill and ability of other pilots, regardless of their race, ethnic, or religious background. There are, however, exceptions.

I recall a young instructor pilot who was desperately anxious to get on with a major air carrier. Every time a woman enrolled in the flight school where he taught, this guy complained bitterly, just knowing that because of some affirmative action program she would wind up "getting his job." It didn't happen because he is now a captain for a major carrier. He also bitterly resented every African-American pilot he encountered because he was certain that because of affirmative action, he would be deprived of the job he wanted.

Then there was the case of the fine young African-American pilot who trained at our flight school. He, too, wanted to get on with a major carrier, and I recommended that he put his picture on his application. In this case, an affirmative action program, all other things being equal, would help him along the way. He refused to do so, claiming that he wanted to be hired for his ability, not his race, a noble attitude, but I still think he should have taken advantage of everything available to him. Again, it didn't matter because he, too, is now flying for a major carrier.

We once had a militant feminist student with a chip on her shoulder. I really believe this woman was paranoid. She blamed every single mistake she made on the fact that because she was a woman, the entire aviation community was conspiring to see that she didn't make it through flight training. She consistently goofed up controller instructions, and then wanted to argue about how the controller was wrong, even when confronted with a tape transcription of the communications. In fact, she even argued with flight service specialists when getting a briefing if she didn't like what she was being told. She claimed they were giving her a hard time because she was a woman. During the course of her primary training, this woman changed flight schools (and instructors within the same school) no less than eight times, each time claiming that she was being discriminated against because of her sex.

A few years ago a young African-American with a pilot certificate came to me for additional training. His certificate read "Commercial Pilot, Rotorcraft-Helicopter, Instrument-Helicopter, Private Privileges, Airplane, Single Engine Land." He wanted to acquire commercial privileges in airplanes and an airplane-multiengine rating. Over a fairly brief period he completed his training for the desired certification. His ambition was to become an air-carrier pilot without flight instruction along the way.

Nearby was a cargo-hauling operation that used Douglas DC3s, and from time to time this company would run a class on the airplane and the company operations for prospective copilots (or first officers, as they are now called), and my student went through that training, finishing high in the class of eight. This class was taught by a former student of mine. I then took him by the hand and introduced him to the chief pilot for the cargo operation, a man who had hired several people in the past based on my recommendation, and who at that time owed me a favor.

My guy was the only one who went through that class of company trainees who didn't get hired. However, it turned out just as well because he got on with a commuter immediately thereafter and is now a first officer for a major carrier. The loser in that situation was the cargo hauler who didn't hire the guy I recommended. They missed out on having an outstanding pilot on their staff.

I recently received a message from a woman who is a senior in a college aviation program. She said, "Just wanted to relate my private checkride experience. I was there for eight hours....She wanted to know everything about that plane. The flight test was two hours long also. I did every maneuver in the book, but this woman was relentless. She was trying to fail me on purpose. I was angry at the time, but being a female in a guy's world, I see that she was also preparing me for dealing with the 'good ol' boy' network....Those of us that are female must do it better to succeed...."

If it is true that this was the examiner's reasoning behind the grueling flight test, there's something wrong with the examiner's thinking. Possibly the examiner was merely probing to make sure the applicant really knew what she was talking about. I know as an examiner, when an applicant gave me an ambiguous answer I kept working around the area until I was satisfied that he/she really understood the material. On occasion, an applicant would answer a question correctly, then keep on talking until he/she talked him or herself out of it. Likewise, in the air if a maneuver was marginal I would have the applicant perform another task with the same objective, and another, and perhaps another, until I was satisfied that the objective of that pilot operation had been met. I'm sure other examiners apply the same technique to evaluation.

I fail to comprehend the concern of those people who worry about this sort of thing. There's really no difference between male and female pilots per se. There are good and bad men who fly (or, rather men who fly well and those who fly badly), just as there are good and bad female pilots. What's the big deal? If one does a good job, it matters not whether you are male, female, black, white, Asian, Catholic, Protestant, Jew, or whatever. And I honestly believe that the majority of designated pilot examiners hold the same beliefs. If not, I believe they can set aside their prejudices when administering a practical test for certification, although I do know a former FAA inspector who is prejudiced against women and African Americans in aviation and who lets it influence his judgment when administering a flight test.

The bottom line is: there's just no place in aviation for prejudice!

## II The FAA revisited

In Chapter 13 I related a few horror stories regarding the manner in which the friendly feds treat the users of the airspace—the pilots. Now if you can stand it, I'll give you a couple of more.

A few years ago, before the United States adopted the ICAO airspace classification system (when we still had TCAs), a very prominent pilot took off from an airport in Southeastern Michigan for Atlanta,

Georgia. He departed VFR in a Beech Model 18 at about 1:00 A.M. on a clear night with the moon out and the sky full of bright stars. He called Detroit Approach Control and announced that he was off PTK (Pontiac) and climbing to 7,500 en route ATL. The controller gave him a discrete squawk code and confirmed that he was in radar contact. He then vectored the pilot right through the DTW TCA (Detroit Terminal Control Area) on the climb. The pilot then proceeded on course for ATL.

The next day a supervisor at the facility was monitoring the tape of the previous night's activity, and when he listened to the conversation between the pilot and the controller, it dawned on him that the pilot had been in the TCA without a clearance, so he wrote it up and turned it over to the Flight Standards District Office for possible follow-up disciplinary action.

The first the pilot knew that anything was amiss was when he received a LOI (letter of investigation). By that time the 10-day limit had passed for filing a NASA (an ASRS—Aviation Safety Reporting System) report, so the pilot retained the services of an attorney and responded to the LOI by requesting an informal hearing. This occurred at a time when the informal hearing was really a total waste of time, for the Associate Regional Counsel had no authority to do anything other than impose the required sanction, which at that time was a 30-day suspension of the pilot's certificate.

Thereafter, the pilot—through his attorney—requested a formal hearing before an ALJ (administrative law judge), appealing the 30-day suspension. At this hearing the controller who had vectored the pilot into the TCA testified under oath that he had meant to clear the pilot through the TCA but had somehow failed to use the magic words, "Cleared to enter the TCA."

After hearing this testimony, listening to the tape, and the FAA lawyer say, "I didn't hear anybody on that tape say the pilot was cleared to enter the TCA," the ALJ made his ruling. He said that the incursion was obviously inadvertent, the pilot obviously thought he was doing the right thing, but even so, he was in the TCA without a clearance and he simply couldn't let it go. He then knocked the 30-day suspension down to 7 days and told the pilot to check back and find a period within the last year or two when he hadn't flown for 7 days and that would be his suspension. In effect, this let the pilot clear off the hook, except for the fact that he now has a violation and suspension on his record.

While this case was working its way up to the ALJ hearing, I was involved in a similar situation. Just by happenstance Atlanta was also involved in my case. I was returning from ATL late at night, VFR at 8,500. It was another beautiful, clear night. As I came into the area covered by DTW approach, I called, advised my type, altitude, and position and requested a descent through the TCA for Pontiac. The controller gave me a discrete squawk and said, "Descend and maintain 3,000."

I pulled the plug and started down. Suddenly I remembered the ongoing case of the TCA violation, so I very deliberately said, "Am I cleared to enter the TCA?"

The controller responded, "Uh...uh, yeah, you're cleared through the TCA." It was quite obvious that he had merely forgotten to use the magic word (cleared). No doubt, nothing would ever have come of this had I not insisted that he issue the clearance. He certainly would not have reported it, but who knows who else might be listening?

Now, many years later, I received today the following e-mail communication from a former student who was flying from PTK (Pontiac, Michigan) to MCI (Kansas City International) via MLI (Moline, Illinois):

*Subj: My latest ASRS report*

*Date: 96-08-26*

*FIRST DRAFT*

*On 8-23-96 I was flying VFR from PTK to MLI. I Flew direct from PTK to Michigan City, Indiana. From Michigan City I Flew direct to MLI. The primary mode of navigation was Loran C with VOR and confirmation by ground reference as supporting information. I was cruising at 4,500 feet MSL. I had been using flight following for the entire route. Near Gary, Indiana, I was told to contact Chicago Approach on 118.4. I gave the controller my destination, route of flight, and altitude. I was given a new code to squawk for flight advisories. At no time was I told to remain clear of class B airspace, nor was I given clearance to enter class B airspace. I wondered about the fact that the controller never mentioned the class B airspace but the controller was busy and had seemed flip on my initial contact and request for flight following. I elected not to irritate the controller by making him say the specific words, "Cleared into Class B airspace." This seemed unnecessary since the controller knew my destination, route of flight, and altitude and had all that information on his radar screen in front of him. My past experience has always been that if the controller wants me clear of class B airspace, they are very specific about routing and altitudes. I expected this controller to follow suit with either passage through or vectors around. At about 2200 Zulu the controller asked if I had DME.*

*I informed him that I had Loran C. He asked me to tell him what my DME was from ORD. I complied and reported my distance. I was just within the southernmost edge of the ORD class B airspace. He asked me to confirm that I was within the class B airspace, which I did.*

*Whereupon he informed me that I required a clearance to be there and I had not received one. I explained that was why I had been talking to him. I asked what he wanted me to do (wanting first of all to correct the situation and discuss the issues later). He replied that there was no problem, that he had me in radar contact, that there were no traffic conflicts, and to just continue on my route of flight. Next I was handed off to ORD Approach, whereupon my route of flight was altered to leave class B airspace.*

*I do not understand why this incident occurred. It is absolutely my responsibility, as pilot in command, to remain clear of the class B airspace unless I have a clearance. Clearly the controller saw the situation develop. If he had a problem with my route of flight, then why did he not address the issue in a timely fashion, on my initial contact, for instance? I had no desire to do anything other than traverse that area of my flight without creating or experiencing any difficulties whatever. I was willing to cooperate in any way the controller had wanted, change my altitude, my route, or whatever was necessary.*

*I feel like I was set up by the controller. The situation was entirely avoidable. I had no desire to be in class B airspace if the controllers did not want me there. The lesson I learned is to have less trust for controllers. I will, in the future, make certain that the controller says, "Cleared into Class B airspace." before I enter.*

This ended the initial draft of the "NASA Report" which the pilot submitted. He also appended the following note to me:

> *Howard: Any thoughts? I feel like the new kid in high school who gets sent out for a bucket of steam. I feel like that guy really took advantage of me, set me right up. For what purpose? Like he wanted to teach me a lesson or something. Beats me. He didn't sound like he was going to write me up by the conversation and tone and all. In fact, at that point he seemed to be trying to reassure me, but you never know.*

Indeed, you never know. I recall an instance in which a pilot—after being counseled by the office manager of his local FSDO—was told that, "You may consider the matter closed. You'll hear no more about it." And four and one-half years later the same manager at the same Flight Standards District Office brought it up and used it against the pilot in another enforcement action. You never know. Is it any wonder that the FAA enjoys so little trust? They have proven time and again that they can't be trusted!

In the case of the letter writer, I told him the story of the pilot who got vectored into the TCA and then charged with a violation for being in the TCA without a clearance, and I advised him to send in the ASRS Report just as he wrote it. No doubt he will never hear from the FAA regarding his penetration of the Class B airspace, but at least he is covered if he should.

In his case it is most likely that nothing will ever come of his "transgression." But if, perchance, the controller does opt to make an issue of the situation, the pilot will be protected against whatever sanction (certificate action or civil penalty) might otherwise be imposed.

Remember, the FAA is in love with words, and it behooves us to make sure both we and they use the right ones, such as "cleared to enter."

# III  Ab initio training

Recently some foreign air carriers have instituted a program of training their pilots (after having them undergo a thorough program of testing for aptitude) from scratch in the sophisticated aircraft they will be operating. I'm talking about major air-carrier transport category jets. Of course, this initial training is accomplished in a simulator. The jury is still out on this system of training, but I recently found an interesting post on the Internet from a lady who started flying at the age of 13 years, who is a flight instructor at a small flight school in the Midwest. Her name is Michelle Panabecker-Neff. Here is her letter just as she wrote it, grammatical errors and all:

> *"From the beginning the budding systems manager is thrust into the world of the glass cockpits, the hubbub of busy airports, and the activity of line pilot duties. The saga (sic) of real world flight environment in the theory that, the established training methods of, gaining a glider rating, an powered aircraft rating as a private pilot, then instrument pilot, then commercial pilot, adding a sea plane rating, or some other experience such as aerobatic's and air combat maneuvering and an Instrument Flight, and ground instructor has no place in to days (sic) computerized aircraft; in this analogy I will explore that fallacy of ab initio training in that it inadequately prepares a pilot to deal with real world flight conditions that do not appear in text books and training syllabus manuals.*

> *It is a fallacy to believe that the small flight school has no place in the development of a professional pilot. (Note: Here she is equating "professional pilot" with "air carrier" pilot.)*

> *Two pilots developed in two different ways, one of them a "A" student in high school, graduated with all academic honors, entered a military academy after graduation, the other a "C" student, at 12 years of age built model aircraft, experimented with wing designed (sic), velocity, lifting bodies, obtained a glider rating at 14 (sic), and was totally immersed in flying — bored to death with the academia in high school, and sought the challenge of understanding the flight environment.*

> *The first pilot, being the outstanding school student; obtained honors at the Military academy, entered flight training in the ab initio programs after all that person tested well, always had the text book answer, made all the right moves in the simulator. The other, they entered into heated debates with their professors, argued about some of the established explanations for air crashes, became a bush pilot because they just weren't what the Military was looking for.*

> *The first pilot, after graduating from the Military's ab initio training program became a Captain in the United States Air*

*Force, in 1996 died in an air crash shortly after takeoff, from a Mountain valley airport, killing the entire crew. The problem failure to recognize when a change was necessary, a departure from the norm due to the existence of a mountain wave. The other pilot, accumulating some 4,500 hours of instructor time; has graduated a few hundred fledgling aviators and has never been involved in and (sic) accident, and neither has any of her students. Many of them are now flying for the Military in combat pilot positions, and the airlines. She maintains that the problem with the ab initio program is it takes time to learn to be an adult! She flies where Eagles dare to walk; but she's just a bush pilot.*

Ms. Panabecker-Neff, who wrote this, appears to hold the mistaken impression that the United States Air Force pilots are trained ab initio in military jet aircraft. She solicited comments in response to this letter and a raging controversy ensued. I don't know how you feel, but I have always been big on OTJ (on-the-job training). For example, I would much rather have aircraft mechanics who trained by apprenticing themselves to old pros while studying for the written do the work on the airplanes I fly than mechanics who got most of their training in the classroom environment with a lot less hands-on, practical work. It seems to me that ab initio pilot training in sophisticated aircraft, with all the simulator emergency work involved, would better prepare air-carrier pilots for their job than several hundred hours churning around the pattern preventing primary students from committing suicide as they learned to fly light training aircraft. The use of sophisticated simulators that faithfully duplicate the cockpit environment and the performance of specific make and model aircraft permit training for emergencies that would be impossible in flight in a real airplane. There is an enormous advantage in this kind of simulator training. In addition to the emergency work which is impossible in an airplane in flight, it is possible for the instructor to stop the simulator at any time and quietly discuss a given situation with the student.

Air-carrier flying has almost no relationship to flying light general-aviation aircraft. Ab initio training teaches the air-carrier pilot how to be an equipment manager (which is what they are), while conventional pilot training teaches the student how to manipulate an airplane (among other things), which is what he or she does. The sort of quasi on-the-job training that the ab initio students get prepares them for what they will be doing. Conventional training does not.

# IV  Commercial operations

In the collective mind of the American public, the term "commercial pilot" is synonymous with "air-carrier pilot," but those of us involved in aviation know better. We know, for example, that there are a great many other kinds of commercial activity in which those of us who fly may rightfully be called commercial pilots. And I don't include all those who hold airperson certificates declaring them to be "commercial pilot." I, for example, hold a certificate which states that I am an "airline transport pilot," but I have never been involved in operations which required such certification, nor do I ever intend to be. (I take that back. I was once the chief flight instructor at a Part 141 flight school, and the FAA required that the chief instructor on the approved ATP curriculum hold an ATP certificate.)

No, what I referred to above as commercial activity involves any piloting activity for which an individual is paid. And this does require the pilot to hold at least a commercial pilot certificate. In this chapter we will consider a few of the many things pilots with commercial certificates (and in most cases, second-class medical certificates) can get paid for doing and thus be entitled to call themselves a "commercial pilot."

## The commercial certificate

When they acquire the commercial certificate, many pilots think this authorizes them to do anything in an airplane and get paid for flying. Not so. Many others think they are authorized only to give "air rides" under day VFR within 25 miles of the departure airport and return to land only at that airport. Also not so. The reality lies somewhere in between these two extremes. The world of flying for hire is quite complex and somewhat confusing. This confusion results from the complexity of those regulations dealing with what the FAA defines as "the carriage of persons or property for hire."

All flying activity is governed by Part 91 of the Federal Aviation Regulations, entitled "General Operating and Flight Rules," but those regulations governing the carriage of persons or property for hire expand on and add to the general rules. The part of the regulations most concerned with the general-aviation pilot is number 135, entitled Air Taxi Operators and Commercial Operators. To fly and be paid to do so under this part of the regs, the operator for whom the pilot works must hold an air-carrier operating certificate, and the pilot must be named under that certificate. The two most important factors required for an operation to fall under this part of the regulations are: one, a "holding out" (being available to the public with an airplane and a pilot), and two, "operational control" (the scheduling and assignment of airplanes and pilots to specific trips).

Other than this kind of activity, there are a host of things that pilots with commercial pilot certificates and current second-class medical certificates may do to earn money by flying. If they hold instructor certificates, they can be paid for teaching, flying as corporate pilots, providing pilot services in an airplane owned (or controlled) by the customer, tow gliders or banners, do aerial application work and aerial photography, and many other things. But what they cannot do is be available with an airplane to carry persons or property for hire.

When administering a certification practical test for the commercial pilot certificate, I used to set up two hypothetical situations. In the first situation I would explain to the applicant that Nancy, the wife of his friend Bill, called the applicant in the middle of the night and informed him that Bill was in an automobile accident in a distant city, was in the hospital and wanted Nancy at his bedside when he went in for surgery. She then asked the applicant if he would be kind enough to meet her at the airport and fly her to Bill's location in the airplane the applicant owned (or otherwise controlled), adding that she'd be happy to pay for this service. I would then ask the applicant if this was okay. The answer is no. To perform this service the airplane would have to be listed under a 135 operating certificate and the pilot (applicant) would have to be named under the certificate.

I would then set up the same set of facts, except this time I would have Nancy tell the applicant that when Bill called her, he asked her to call the applicant and ask him to fly Nancy to the destination in Bill's airplane (or the airplane controlled by Bill), and offer to pay the applicant for this service. Is this okay? In this case, the answer is yes because under this set of facts he is providing pilot service in a customer-owned or controlled airplane. If the applicant clearly understood this difference, it was a good indication that he or she had a basic understanding of the privileges and limitations of the commercial pilot certificate.

The FAA's definition of "commercial operation" is indeed unique. For the FAA to consider it a commercial operation, somebody has to be being paid to sit in the aircraft! For example, airplanes used by a flight school — if they are used for dual instruction and an instructor is being paid to teach in them — are considered to be in commercial use and are required to undergo 100-hour inspections. However, those same airplanes when rented for solo flight by students or to certificated renter pilots are not considered to be in commercial use and thus are not required to undergo 100-hour inspections. It doesn't seem right, but that's the way it is.

## The corporate structure

Many, if not most, large corporations that do business over a substantial geographic territory maintain large and somewhat elaborate flight departments for the transportation of their executives and others who are required to travel extensively to further the company's business interests. These corporate operations may have large fleets of transport-category aircraft and run regular schedules between far-flung branches of the company. An example of this kind of flight operation would be the big three automobile manufacturers. And their pilots are paid salaries commensurate with those of the major air carriers. At the other extreme is the company that owns (or leases) a single airplane and employs no pilots at all but calls in free-lance pilots as needed. (Or it is flown by a company employee, usually executive type, whose primary function is something other than pilot.)

There is also a large group of corporations that share ownership of one or a fleet of aircraft. The usual procedure in this case is for an FBO (fixed-base operator) to manage the aircraft and supply the crew. This kind of operation may fall under either Part 91 or Part 135 of the regulations, depending on the interpretation of the local FSDO and on whether or not the FBO owns a share of the airplane or fleet of airplanes.

And there is a large and growing group of companies that own one or more airplanes but do not have pilots on the payroll. These companies have their equipment crewed by the personnel of an FBO that manages the aircraft, or by commercial pilots who are in the business of providing "pilot service" to the ownership of such aircraft. Whatever form the operation takes, corporate aviation provides a substantial amount of employment for the commercial pilot population.

There are a few (a very few, but growing number of companies) that have one full-time, salaried pilot on staff and who "sell" right-seat copilot time to pilots who are willing to pay for the experience gained by flying in heavy equipment or jet aircraft. If the captain happens to be a certified flight instructor, these experience seekers can actually get training for their money.

## Air charter
### The cargo haulers

Again, in the collective mind of the American public, air charter means on-demand passenger carriage. But at least as important as passenger carrying is the vast amount of cargo hauling by air that goes on. In this kind of activity also there is a huge disparity between the largest and the smallest operators. At the top are the regularly scheduled freight carriers flying transport category jets such as FEDEX (Federal Express) and UPS (United Parcel Service). These operations have large fleets of aircraft, and their pilots are very well paid. Then there are the large on-demand freight haulers that also use transport category aircraft. They, too, usually have substantial fleets of aircraft. They stand ready to move fast freight all over the country on very short notice.

Although the electronic transfer of funds is a fact of life in today's business world, paper (checks, currency, drafts, etc.) is still moved by air. All night, every night of the business week, the skies are full of airplanes carrying canceled checks on their way to the Fed (Federal Reserve Clearinghouse).

The world of cargo flying goes all the way down to the one-person, one-plane operation which may serve only a few, or as little as a single, customer—any kind of business that occasionally requires fast freight delivery. These kinds of operations can be found around any major population center, and they fill a definite need in the transportation system. Here, too, we are beginning to see a situation where some operators are charging eager young pilots for the privilege of providing copilot service while they gain experience.

## Passenger charter

As is the case with the cargo haulers, on-demand passenger charter runs the gamut from large operations with a fleet of multiengine, or even jet-powered, aircraft to the one-person, one-plane operator. Many companies do both passenger and cargo charter, but these are usually the smaller charter outfits. And many others, such as flight schools, whose primary business is something else (training), do a bit of charter work on the side. And, of course, there is quite a bit of specialty work in this area as well—for instance, the entire realm of air ambulance flying, using both fixed- and rotary-wing aircraft.

The short-haul commuter airlines have traditionally been faced with the problem of investing huge sums of money in training their crews, only to lose them as soon as they can get on with one or another of the major carriers. These commuter airlines offer absolutely abysmal pay and virtually all their pilots are just waiting to get hired by a major air carrier. Several means have been worked out for the companies to recoup the investment in pilot training, including requiring the pilots to pay for their own training (they call it "buying a job") and holding back a portion of their pay until they have remained on the job long enough to justify the company's investment. Now we are beginning to see the same situation described above in which the pilots pay for the privilege of gaining experience as copilots. No matter how it is done, this situation of low (or no) pay for the pilots coupled with the extreme cost to the company of crew training has been a bad deal all around. However, there is nothing in the regulations that requires that a commercial pilot be paid.

## Specialty flying

As a guy who employed pilots, I always gave extra consideration to those with specialty flying in their background. If pilots had been hauling cargo or flying the mail or canceled checks all night in all kinds of weather in tired old Beechcraft Model 18 airplanes, if they had been towing banners or gliders, I could feel confident in their abilities to make accurate decisions. Both bush pilots and crop dusters are each a breed apart. These two kinds of commercial activity are indeed unique and require very specialized skills, unlike anything else. Aerial survey and photography flying is another kind of commercial specialty requiring unique skills and additional training. Pipeline patrol and fish and wildlife spotting are also considered specialty flying in the world of commercial aviation. The pilots who do this sort of work are very likely to be truly expert aircraft manipulators.

Most towns or cities of any size have radio or television stations which employ airborne traffic reporters and which employ pilots to fly them around. And, of course, there are the law-enforcement pilots, including those who are federal agents (FBI, DEA, Customs and

Immigration, etc.), state police, county sheriffs' deputies, and local police. Each of these specialties requires unique piloting skills. All of these people and the professional flight instructor are truly "commercial pilots," so you see it is not just air-carrier pilots who are entitled to be called commercial pilots, but anyone who earns his or her living by flying.

## The professional flight instructor

Flight instructors who are imbued with the zeal to help their students—whose desire is to see that every minute that students are paying for their time is a valuable learning experience—these instructors are truly to be admired. They have spent untold dollars and countless hours perfecting their teaching skills, and it shows in the way in which they comport themselves. Such instructors are well worth whatever their fees are (and it is no doubt not enough). When compared to others who have invested far less time and money in their preparation, the professional flight instructors' earnings are pitiful indeed. And this is true whether they are full-time professionals or part timers who merely want to share their love of aviation with their students.

Unfortunately, flight instructors occupy the very bottom of the ladder in the ranks of commercial pilots. This is because instruction has traditionally been the avenue that individuals have had to travel on the way to a "real job" in aviation. It is often called "paying one's dues." With the advent of ab initio training for air-carrier pilots, we are currently seeing a slight, ever so slight, change in the system. We can only hope that this concept will grow and that professional flight instructors will begin to occupy their rightful place in the scheme of things.

For me, personally, primary instruction is the most gratifying of all. I enjoy teaching, and there's nothing I know of that one can teach in which he/she gets as dramatic a demonstration of the results of his or her effort as is the case with primary flight instruction. I've taught huge classes in a large university, and once in a while I'd get to see that rare spark that lets me know I was reaching a few students, but in teaching beginning flight students, when you turn the key that opens their eyes and helps them over one of the classic "learning plateaus," the students literally light up. "Look! It works! It works!" they exclaim. I eat that up. Nothing that I know of comes close to matching this for job satisfaction.

## The pinch-hitter program

Probably the most single gratifying experience of my long career as a flight instructor occurred several years ago when I had a woman pinch-hitter student. Our local chapter of the Ninety-nines International, the worldwide women's pilot organization, annually sponsors a Pinch Hitter course for the frequent passenger, and I had been volunteering a weekend each year for several years to teach in this program. Pinch-hitter students, sitting in the right seats, get 4 hours of flight training, during which time it is hoped they will learn enough to navigate to an airport and get the airplane on the ground without hurting themselves or anyone else or seriously bending the metal.

On this particular occasion I had the wife and daughter of a man who owned a Stinson Station Wagon, a particularly fine specimen of an old taildragger. If the young daughter had been old enough and had a student pilot certificate, I believe I could have soloed her — she was that good. Her mother, however, was quite a different story. The poor woman was so nervous that the first time we went up, she folded her arms across her chest, closed her eyes, and flatly refused to touch anything. I abbreviated the session, flew a tight pattern, and landed. The amazing thing is that by the end of the second day, after our fourth hour together in the air, I had that nice lady landing the airplane completely unassisted! I am convinced that in an emergency if she is flying with her husband and he should become incapacitated, she could find an airport and land the airplane. She might run off the runway, but nobody would be injured and the airplane would be relatively undamaged. You can well imagine the sense of accomplishment I have felt as a result of that experience.

This pinch-hitter program, by the way, is one in which I believe quite strongly. Over the years it has been responsible for numerous "saves" in which a nonpilot passenger has successfully brought an airplane in to a safe landing when the pilot became incapacitated during the course of a flight. The curriculum was designed by the AOPA (Aircraft Owner's and Pilots Association) and it includes four hours of ground instruction, four hours of flight instruction, and four hours of after-flight debriefing. In the program offered by our local chapter of the Ninety-nines, the ground instruction is administered by members of the Ninety-nines who hold ground instructor certificates, and the debriefings are conducted by Ninety-nines who are familiar with the specific make and model of airplane flown by the pinch hitter. Of course, the flight instruction is given by certificated flight instructors, volunteers from local flight schools, or free-lance instructors from the area. Many of the frequent passengers who go through the Pinch Hitter program go on to undertake formal flight instruction and become certificated pilots.

# 16

## More of what comes next

### I The road to the left seat in a major air-carrier airplane

If your ambition in aviation is to be a pilot for a major air carrier, you are a member of an enormous group of people. It seems that this is the burning desire of many, if not a majority, of today's beginning pilots. Perhaps it is the really large salaries that attract them. Or maybe it is the glamour of the job, in which case they are sadly mistaken, for there is nothing glamorous about being a glorified bus driver. But whatever their motivation, hordes of young pilots are simply dying to get on with the carriers, so we are devoting a not inconsiderable amount of space to describing the means by which they do so.

For many years the major air carriers recruited their pilots from the military, and to an extent that is still the source of a fair portion of the heavy-iron pilots. The downsizing of the military in general has resulted in there being fewer military pilots, and this, coupled with the inducements the government has been offering to keep its pilots, has resulted in the carriers having to look elsewhere for most of their recent hires. Here, however, we will deal only with the path followed by civilian-trained pilots.

Traditionally, the usual path followed by the civilian-trained pilot to the airlines has been to first instruct for a year or so, become a charter pilot or freight hauler for another couple of years, accumulating multiengine time, then fly for a commuter for a while, building turbine time, and then to finally get hired by a major air carrier.

Sometimes after instructing for a year or so, pilots will go to work for corporations and build their experience as corporate pilots, then go directly to the airlines, skipping the charter and commuter stages. However, with the phenomenal growth of the commuters since deregulation, it is here that most of the recruiting is done by the majors. As the majors gave up their unprofitable routes, the regionals sprang up to fill the gap. The commuters had been flying twin Cessnas and Navajoes, then they began to use turboprops and the smaller jets, providing a trained pool of pilots for the major air carriers.

This situation posed a very real problem for the regionals. With the majors experiencing a shortage of pilots, they recruited from the regionals, who found themselves in the position of having invested a substantial amount of money in training pilots, only to see them jump to the majors in a very short time, often in a matter of only a few months, and long before their employers could begin to recoup the investment they had made in training the pilots. This situation has resulted in several factors which impact negatively on the career of the pilot hoping to fly an airliner.

First, the simple economics of trying to stay afloat in the regional airline business necessitates that the pay scale of their employees be abysmally low, and the pilots frequently have the added duties of being ticket agents and baggage handlers as well. And second, we are now seeing a situation in which commuter pilots have to pay for their own training, and we're not talking about basic training, but airline-specific training after having a job commitment, or maybe even without a job commitment. More and more, pilots are having to pay substantial fees just to submit applications to the regionals. Then, if selected, they are required to pay upwards of $10,000 for training, usually without even a firm job commitment. No wonder it's called "buying a job."

There are also "intern" programs sponsored by both some regionals and even some majors. These programs work through a couple of the large training academies, and here again there is no guarantee of a job after graduation. The entire job situation is a sort of "filter-down" process. If and when the majors experience a pilot shortage, they raid the regionals, and job openings suddenly appear at that level. When there is a glut of pilots and the majors aren't hiring, as is usually the case, the crews manning the regionals are stuck there. Traditionally, no pilot has expected to make a career of flying for a regional airline, but now, despite the extremely low pay, we are seeing for the first time a situation in which some pilots are doing just that. It is a market-driven situation, a simple matter of supply and demand.

## Three case histories

Several years ago I hired a young instructor who taught at my flight school for about a year, earning in the mid twenties. He left that position to take a job flying canceled checks at night. He started that one with an income in the mid teens and worked his way back up to the low twenties in the course of a year or so. He left that charter company to fly for a regional, this time starting in the very low teens. After a couple of years he was back up to the high teens or low twenties, when he got hired by a major carrier, starting as a first officer, again dropping back to the mid teens as a starting salary. He's now a captain, earning a salary well over six figures. Did he pay his dues? You better believe he did! His story is fairly typical of the road followed by those who wind up in the left seat of a major airliner.

More recently a young man came to me for training. He had a commercial pilot certificate for rotorcraft-helicopter, instrument-helicopter, and private privileges, airplane-single engine land. He was (and still is) a National Guard helicopter pilot. He wanted to acquire an instrument-airplane rating, commercial privileges, and a multiengine rating in airplanes so that he could qualify for an airline job. He had no desire to instruct, and as soon as he completed the training for commercial, multiengine, and instrument fixed-wing ratings, he went to work flying canceled checks at night. He then flew fast freight in a twin-engine airplane as a contract pilot for a large multinational corporation for about a year and a half.

He left that position to work as a commuter pilot flying a turbo-prop, first as a first officer, and then as captain. He is now a first officer for a major carrier. Again, he is an example of one who paid his dues and realized his dream. And again it wasn't an easy road to travel.

More recently, a friend of mine, a free-lance flight instructor, trained a friend of his for an instrument rating, using one of our school airplanes. I administered the practical test for the student's instrument rating. He was employed as a computer programmer, earning in excess of $50,000 dollars per year, but his driving ambition is to be a pilot for a major air carrier. My friend, the instructor, asked me to help the young man along his way, so I trained him for a multiengine rating, and then a commercial certificate, airplane-multiengine land, private privileges airplane-single engine land.

He did not want to instruct, but he felt that he would have to do that to build enough time to make himself attractive even to the regionals, which he believes is the last step on the way to the majors. While I was training him for single-engine commercial privileges, going through the commercial maneuvers (chandelles and lazy eights), he remarked, "I hate this stuff!" Well, I certainly didn't want this guy to be an instructor. It was obvious if he didn't enjoy chandelles and lazy eights (most pilots think these are pure fun), and if he hated spins (as he said he did), he was bound to be an absolutely terrible instructor. He wouldn't want to do it, and it would show right through to his students.

Even so, I went forward with his CFI training, knowing that if I didn't, somebody else would be happy to take his money and not train him, but I would (I think I'm an outstanding instructor). However, I was desperately anxious for him to not become an instructor. We got lucky. After he got his multiengine rating, several opportunities opened up for him to gain experience in twins. (I had him make a couple of airplane deliveries that required trip lengths of over a thousand miles.) He recently landed a job as a copilot flying heavy iron for a cargo-hauling outfit. In a very few years, after being type rated in several transport category airplanes, he will be ready for the majors. And the good part is he'll never have to instruct. He will, however, be required to suffer a severe financial sacrifice, since he quit his high-paying computer job to do what he's now doing.

Obviously the path to the left seat of an airliner is a rocky one, strewn with obstacles, but not an impossible one for the truly determined individual. And neither of these two individuals had to "buy their jobs."

# II Aviation and the media

## "You're lucky to be alive"

At my age, I suppose I could very well get up every morning, look in the mirror, and if I'm there, tell myself, "You're lucky to be alive!" Given a choice between luck and skill, I would rather be lucky than skillful. However, it has been said that skillful pilots make their own luck. But that's not what this is all about.

Several years ago, due to a mechanical failure in the throttle linkage, I had to put an airplane down in a field about 1½ miles from the airport. (I was on an approach, and after descending to the MDA (minimum descent altitude), I advanced the throttle to arrest the descent and got no response.) The off-airport landing was strictly routine— no damage, no injury to either my student or myself.

About an hour after I had been picked up and returned to my office at the airport, a reporter from the local (sort of small-town daily) newspaper stuck his nose in the door of our large, open office and asked, "Where's the pilot that crashed down the road?" The office manager pointed to where I was sitting at a desk across the room, and the reporter extended his index finger, and pointing to me, shouted, "You're lucky to be alive!"

I admit I made the wrong response. Instead of explaining that it was a very normal, routine landing, I threw the fellow out. Stay with me a while and you'll discover what I should have done. In case you're interested, the problem that time was the throttle linkage. It had come off at the front end and the engine went to idle and stayed there. After temporary repairs, I taxied the airplane down the road to the airport. (With a police escort!)

Ever since the beginning of manned flight, the popular media have grossly misrepresented aviation in general and general aviation in particular. And the result of this misrepresentation has been a very deleterious effect on the entire aviation community in several different ways. The friendly feds (the FAA), which are charged with the responsibility of "Fostering and encouraging the development and growth of civil aviation in the United States," instead of countering adverse publicity with facts and knowledge, cave in and pass new regulations. And the innate fear of the unknown has caused countless people to shy away from flying, even as passengers on air carriers, the safest form of transportation known. Airplanes are an easy target and the media is quick to seize upon any sensational event.

Dick Knapinski, who handles public relations for the EAA (Experimental Aircraft Association) in Oshkosh, Wisconsin, has this to say:

> *On a year-round basis, we hear from a wide variety of reporters regarding aviation. Many of these inquiries, unfortunately, come after an accident—usually fatal—and the reporters are bumping up against a deadline. That perhaps is one of the key instigators of general media goof-ups. Daily reporters, unlike their magazine counterparts, have very little time to put together their stories. They need (quick) facts, not long technical explanations. Nearly all of them have little*

*idea of what an airplane does other than deliver them to Cancun once a year. Small airplanes are cramped, they bump around in the air, they crash for no apparent reason—that's the perception. Many reporters have flown in a small airplane only because they had to for a story and don't have a great personal reference to them.*

## The FAA response to air-carrier disasters

Whenever there is negative publicity regarding aviation, the FAA has responded with new and usually unnecessary regulations. Because of a mistake on the part of a cargo handler, a discount carrier is grounded. As far as I can tell, there was no valid reason for this action on the part of the government. In the normal course of events, the carrier would have been given the opportunity to correct whatever discrepancies were noted, particularly in view of the fact that the causes of the grounding were known to the FAA long before the incident which precipitated the grounding. This is just one recent example of the knee-jerk reaction to a specific event. The list goes on and on. It seems that every mishap, no matter what the cause, results in a great cry of "flying is unsafe" on the part of the mass media. And this, in turn, results in even more unwarranted regulation.

The irresponsible reporting of aviation accidents is based on two factors. The first of these is ignorance, a lack of knowledge regarding aviation. And of equal importance is the fact that sensationalism sells. It generates viewers, readership, and listeners, which in turn creates more advertising, which translates to dollars. Knapinski points out that the first problem is the need to explain a complex issue within a short time frame to a person who doesn't really understand it.

Everything we do involves some degree of risk, and flying is no exception. The safety record of the air carriers is absolutely phenomenal, but because so many people are involved, and the occurrences are so rare, each accident generates a substantial amount of publicity, and the reaction of the FAA is to promulgate a new, and usually unnecessary, regulation. We rarely see this kind of response to other kinds of disasters—although new building codes as a result of earthquake activity do come to mind.

Whenever an isolated terrorist act brings down an airplane, in reaction to the outcry in the press and broadcast media, the government spends more money on useless security measures, virtually all of which do no good whatsoever. Any determined individual can easily circumvent all the security measures now in place or planned for the future no matter how they are designed to work. An individual wearing white coveralls and carrying a couple of covered buckets and a paintbrush could walk right by the metal detectors and guards with firearms or explosives in the buckets. More guards, more background checks of personnel, more metal detectors certainly wouldn't prevent this, nor would similar methods now being used, nor would more security fences around airports. Attempting to limit access to airports and airplanes is therefore futile. At any time by any number of means, anyone can get by these pitiful attempts to limit access.

The only effect these governmental actions really have is to placate the press, and consequently the public, which has led to the false belief that something positive is being done, when in fact all that has been accomplished is the waste of even more money that would be better spent elsewhere.

## General aviation

The situation with respect to general aviation is similar in kind, but even worse because it is greater in degree. This is so because in the collective mind of the general public, light plane flying is an inherently dangerous activity engaged in by a few foolhardy souls who are too dumb to know they are risking their lives every time they take off.

Thus, when the mass media sensationalizes an event, it has the advantage of blowing up a preconceived notion. It fosters the perception that "everybody knows those Piper Cubs are unsafe." Whenever a lightplane goes down, the media scream about how another one of those dangerous things fell out of the sky, killing all aboard (even if the pilot was alone). And of course there is the usual cry to do something to stop this senseless death. I guess what they mean is to ground all general-aviation aircraft. Back to the simplistic answer—if there were no airplanes and no pilots, there would be no aviation accidents.

Because some dumb-dumb flight instructor permits a publicity-hungry father to pressure the instructor to take off in weather so bad the birds are walking, we get proposals for new regulations prohibiting underage children from touching the controls of an air-plane. And given enough publicity, the Congress (never a group to shun publicity) gets in the act, demanding new rules and regula-tions, as if it is possible to legislate stupidity. What the media and the Congress failed to take into account in this case is the fact that there is no such thing as a "seven-year-old pilot." The kid was a passenger! And the FAA, being very sensitive to just how the politi-cal wind is blowing, responds by giving the Congress just what it wants. And another useless (and unenforceable) regulation is the proposed result.

The Jessica Dubroff case presented a perfect example of media frenzy. Literally everyone in the civilized world heard of that one. To begin with, the whole thing was a publicity stunt and was de-signed to attract the attention of the media. And they did such a good job of doing just that, that when the disaster occurred the attention of the media was already focused on the flight of the so-called "seven-year-old pilot," who was attempting a point-less, meaningless nonrecord.

Of this one, Dick Knaplinski says:

> *The best thing that came out of that tragedy is general avi-ation got a rare opportunity to present its case in public. Hundreds of private airplanes take off and land every day, so there's no news in that. We (the EAA) got lots of calls right away, from the local shopper paper to CNN and ABC. We got to discuss kids and flying, flight instructors, flight safety, weather decisions and much more. Of course, in our case, Young Eagles (an EAA program to introduce young people to aviation) was brought up quite a few times. It's not the way I want to get publicity, but it had many reporters call-ing and asking about the program. We got to explain the tight parameters for Young Eagle flights and how that dif-fered from the Dubroff attempt.*
>
> *Both J. Mac McClellan, Editor-in-Chief of FLYING magazine and Phil Boyer, President of the AOPA (Aircraft Owners and Pilots Association), went on national television and were given an opportunity to explain just how that particular pub-licity stunt was by no means typical of what general aviation is like.*

## What to do

If you're a pilot (and I can't imagine why you would be reading this if you're not), you no doubt are among the vast majority of us who love aviation and deplore the bad press we get, and you will want to do whatever you can to improve this situation. I certainly don't have all the answers, but I can offer a few suggestions.

Now that we know what's going on, what's causing all the adverse publicity regarding aviation in general and general aviation in particular, what can we do to change the situation? I believe that for the most part, all this adverse publicity is the result of ignorance rather than malice. (I know of no one in the media who deliberately wants to harm aviation.) And the way to combat ignorance is with knowledge. If we want to improve the sad state of affairs that currently exists between general aviation and the media, it is up to us to embark on a campaign of education. We must convert the members of the popular media from being adversaries to becoming friends. It won't always succeed because sensationalism sells newspapers and attracts listeners and viewers, and this, in turn, generates advertising revenue. But even if our effort results in only a moderate amount of success, it will make a substantial difference in how we are perceived by the public, and this certainly is a desirable objective.

The first step, and any private pilot can do this, is to get the name of the individual who wrote the last story on a general-aviation event in the paper, and the name of the radio reporter who last covered a general-aviation story, and the name of the television newsperson who did so. Then contact each one personally and offer to buy him or her lunch (these people are great freeloaders and it is doubtful if any of them will turn down the offer of a free lunch). Do not become emotional. Remain calm. Be friendly, even subservient, and flatter the person. I guarantee they'll eat it up! Now you are in a position to do some good. Invite them to attend the next aviation function scheduled in your area, FAA safety meeting, EAA meeting, whatever.

Over lunch explain just where you are coming from. Tell them, individually, all the wonderful things that are accomplished by general aviation. Then offer each of them a ride in your airplane. If you don't have an airplane at your disposal, see if you can get a flight school to give the reporters each a flight, or better yet, a sample lesson. If these suggestions are followed in a warm, friendly manner, you should see a positive result. It won't work miracles, but it will pry open some minds.

That is just what we did a few years ago when we had a glider training program at one of the three flight schools my company was running at the time. One of the local television stations sent a camera crew and reporter out for a glider flight, and we got a nice 5-minute segment of their 6:00 o'clock news. We also got one of the weathermen from another television station out for a sample lesson, along with his camera crew, and we gained another 5-minute segment on that station. Any flight school that I know of would be glad to get that kind of publicity, and it is easy to do. All you have to do is ask.

Now that you've made some new friends, offer to give accurate technical advice next time they are assigned to do a story on a general-aviation event. Any time you see an inaccurate story in the paper or see or hear inaccurate reporting on the broadcast media, jump in with a letter, politely offering to set the reporter straight.

In getting these things I've outlined done, enlist the help of all of your pilot friends and the flight school operators and FBOs. Sell them on how helpful this will be to all of us. Why must you be the one to take the initiative in this area, you ask? Because you're the one that's willing to do it, and it must be done by somebody. Otherwise we will continue to suffer the results of bad publicity stemming from ignorance. The great unwashed (the general public) is also extremely ignorant in matters aeronautical, and it's up to us to educate them if we want to be seen in a more favorable light. As one correspondent put it, "Get real. The uninformed (are) uninforming the uninformed. Shouldn't we, the informed, be weighing in? One would think the FAA might be the champions of our cause, (but) they seem more inclined to turn tail when the going gets tough."

When I asked for his comments on the subject of the media and general aviation, Dick Knapinski also had this to say:

> *The annual convention here (at Oshkosh) is ... an interesting situation. With more than one thousand media people here, aviation knowledge runs the gamut of knowing much more than I ever will to not knowing which end of the airplane is the "spinny end." Some of the general media show up at*

*Oshkosh as they would at an auto race—hoping for some crash footage. That's part of the game (and) I can live with that. The best way to frustrate them is to run a super-safe event and make them work for a story.*

*A Milwaukee TV reporter called on opening day, saying that she would be doing a live report at Oshkosh at 5:00 p.m. "No problem," I said. She then asked if she could get some background info. "Sure," I replied. The first question out of her mouth was: "You guys have been known to have some safety problems in past years. What are you doing to make this a safer event?" "I'm not sure what you're alluding to," I said, trying to run through my mental Rolodex of what specifics she might be getting at. "Well, you know, you've had some fatal crashes in the past...." "Ma'am, we haven't had a fatal accident on the grounds since 1981." (I always reflexively start referring to people by sir or ma'am when I feel the "idiot reporter" alarm go off and the heat rise on the back of my neck.)*

*"Could you tell me about that one? There were some others before that, too. I remember when I was working as a college intern here in Milwaukee in the mid-1980s.*

*Could you give me some details about them?" "No, ma'am, I'm afraid I couldn't since I wasn't on staff at that time. Furthermore, I'm not really sure what things from five or ten years ago have to do with opening day today." "I'm not looking for crashes or anything," she insisted. This will just be a small part of the report—five seconds or so." "Well ma'am," I replied, "We seem to have spent an awful lot of time on it for a five-second report. Suffice it to say that safety is our primary concern and FAA, EAA, and local airport officials spend an awful lot of time keeping that high level of safety here each year."*

*She did back down a bit after that, although she did ask Tom Poberezny (president of the EAA) the question later on. Tom is very good at shutting those kinds down in 25 words or less and leaving the impression on the average viewer that the reporter is grasping at something to ask because he/she didn't do any background. Having a president who understands the media beast is a great ally.*

*The worst reporters at the convention, we've found, are those who blow into town (not to pick on TV, but they come to mind more often) and have to do a fast report and just want something now. They don't want explanations of aircraft or aviation. We had one local TV person refer to a B-52 as a "B-52 Stratocaster." That's after he asked for the nickname and wrote it down.*

*The absolute worst media irresponsibility we've encountered is about five years ago when a local TV station reported that a night warehouse fire in another part of Oshkosh was caused by an airplane from the fly-in crashing into it. They never checked that report, or else they would have known that no airplanes fly in Oshkosh at night because the airport is closed. In fact the station never checked with anybody (police, fire, sheriff) but just ran with the story. We leaped all over them for that (again, the importance of calling right away), calling the newsroom, the news director, and station manager, the latter two at home. We did get an on-air apology for that one.*

*We have a should-be-sainted volunteer in press headquarters who answers the phone and regularly gets just one question from the media caller. "Any crashes or incidents today?" She always replies, "We have none scheduled."*

On the subject of dealing with the media for the rest of us who are not public relations specialists, Knapinski has this to say:

*Here's the first problem — the need to explain a complex issue within a short time frame to a person who doesn't really understand it. Reporters, nearly to a person, really try to understand the information you're giving them. It's rare, however, that they get it on the first try. Unfortunately, you often only get one try. That's a challenge for those of us in aviation to boil down our explanations to palatable yet thorough descriptions. An example: there's been a homebuilt accident somewhere. Somebody tells them to find out more about homebuilt aircraft. The title itself is a little scary to a reporter who may have never put anything together more challenging than a bookshelf. Among the first questions are always: "What are homebuilt aircraft?"—It's a category of FAA-approved aircraft that are built by those who fly them. "Can anybody build them?"—Anyone can build one, but the airplane must be inspected and approved by the FAA before it's flown. Pilots are required to have the same training and federal licensing as those who fly the Cessnas, Beeches, or Pipers out at the local airport. I found that relating these airplanes to "real" airplanes works well. "Are they safe?"—Since there really isn't a complete database of homebuilt versus manufactured aircraft yet, our best information is that homebuilts are insured at about the same rates as manufactured aircraft, so that apparently indicates approximately the same risk factor.*

*A danger when talking to a reporter is being dragged into the "what happened?" discussion. Often a reporter will ask without any malice, "So what could have happened?" Human nature wants to supply an answer. But since we weren't in the airplane, we don't know.*

# III  More bad stuff

## Priority again—engine failure

If I had to describe in a single word the one most valuable asset a pilot can possess, I would use the word *anticipate*. In other words, the ability to anticipate, to know what's coming up next (and, as Rod Machado puts it, the next after that) is the most important tool a pilot can have. However, this ability must go beyond that. It must also be coupled with the attribute of flexibility, the ability to change courses of action when it becomes apparent that the present course is not the best choice.

There are really only two situations in normal flying in which it is absolutely essential that the pilot perform a series of tasks in the proper sequence. The first of these was covered in Chapter 3 when we dealt with recovery from unusual attitudes in the cloud, and the second occurs when there is an engine failure in a single-engine airplane, and although almost all instructors teach the proper method of responding to a power loss, a substantial number of pilots somehow manage to permit their emotions to take over and fail to apply what they have learned, perhaps many years earlier.

When an engine quits and the airplane becomes quiet, the airplane becomes a glider, a not very efficient glider, but a glider nonetheless.

When this happens, pilots must immediately latch on to that single, published airspeed that will keep them in the air the longest, even if they have to zoom up a bit to acquire that very specific airspeed. All the pilots have going for them are the few precious seconds of time before they encounter Mother Earth, because surely they will. Go any faster and you'll come down sooner. Go any slower and you'll come down sooner. (Of course, while acquiring that best glide speed, the carb heat should simultaneously be applied if it is called for.)

The most important thing is to keep your cool and fly the airplane. If and when you should have to make an off-airport landing, try to land into the wind if possible, but there are many reasons why that may not be possible, among which are the shape of the landing area selected, obstructions on final, and the direction of furrows in a plowed field. From altitude everything looks flat, but as you descend you are likely to discover that the area you selected isn't flat at all. If it is a plowed field and the furrows are curved, you can bet it is hilly, but if they are in straight lines, it is most likely flat. If it is possible to do so, try to land upwind with the furrows, but it is usually better to land downwind with the furrows, rather than across the furrows.

The key to success is to keep flying the airplane. Don't panic! And don't worry too much about saving the airplane. That's why you buy insurance. Remember, airplanes don't bleed. Do not stall it in. Even if you are landing in trees, it is better to fly it in and let the wings take the impact than to stall it in and have your body suffer the impact. Consequently, if you wish to avoid being the first one to arrive at the scene of the crash, do not permit the airplane to stall!

The next step after latching onto the best glide speed is to look around, pick out a suitable landing area, and start maneuvering toward it. All landings, emergency or otherwise, end up on the ground. Therefore, the best preparation for an emergency landing is to be flying where there is a suitable landing area within gliding distance. Minimize your time and maximize your altitude when flying over unfavorable terrain. When choosing an emergency landing site, there are several factors to consider, among which are the size of the area, the surface condition, approach slope, obstructions, and wind. I know it is asking a lot to evaluate an emergency landing site with all these factors in mind while the emergency is unfolding, particularly if time or altitude are in short supply, but pilots who fly with one eye outside constantly evaluating and asking themselves, "Where would I land if...?" remain mentally ahead of the situation. The advantage of having sufficient time and altitude to adequately evaluate the situation cannot be overemphasized, though many factors may not become apparent until you get closer to the ground. Fences and power lines are often invisible until it is too late to avoid them.

What makes a good emergency landing site? Probably the best is the open field, particularly if freshly plowed, newly sown, or freshly mowed. Good plowed fields will have straight furrows, and though landing into the wind is preferable, never land across the furrows. Curving furrows are signs of hilly countryside. Roads, though generally smooth as a landing surface, present hazards in the form of power lines criss-crossing the road, road signs, and markers which may not be seen until it is too late. The median in a divided highway may be used, but bear in mind that they are usually "V" shaped for drainage.

In any event, plan the approach, into the wind if possible, parallel with the ridges if a plowed field, etc. Then and only then should any thought be given to attempting a restart. There are really only three things necessary to make an internal-combustion engine work: fuel, air, and spark (ignition). These should then be checked. If the pilot's big fat knee knocked off the ignition, it can be turned back on. The application of carb heat has ensured that the engine is getting air. If something on the floor has inadvertently turned the fuel valve to the off position, it can be turned back on (or to the proper tank if the selected tank has run dry). If the restart attempt is successful, fine. Pilots can just keep on truckin' along as they go about their business.

If, on the other hand, the restart attempt is not successful, the airplane will end up rolling out on a nice, smooth surface that the pilot picked out and onto which he or she has controlled the airplane. However, if the pilot wastes time dinking around with the mixture, ignition, fuel, etc. while gently sinking toward the ground at some speed other than the best glide, the pilot is narrowing the radius of action until all that's left is some totally unsuitable landing area such as houses, rocks, trees, etc. and the pilot has permitted the airplane to fly him or her, rather than controlling the airplane to a landing under his or her control.

All this may seem elementary, but I have seen numerous pilots, when given a simulated engine failure, totally ignore the proper sequence as they desperately tried to get the engine going instead of grabbing the best glide speed (even if they have to dissipate some energy by zooming up to get it) and planning an approach into a suitable area.

And while we're on the subject of forced landings due to engine failure, you should know that it is much more common to over-shoot than to undershoot the selected field when an actual off-airport landing is required. Seems that pilots are so careful to avoid undershooting that they keep their approaches much too high to get rid of their altitude (even with the application of full flaps or by slipping it in) that they wind up overshooting and end-ing up in some totally unsuitable area. As an interesting sidelight, one time while administering a private pilot checkride to a young female applicant, I pulled the power and announced that we had lost our engine. Although there was a very nice little airport within easy gliding distance, she selected a plowed field about a quarter mile past the airport, which, by the way, she never saw. She passed anyway since she did everything right, and after gliding right over the airport, she would have made the field she selected. It was a suitable landing area.

A study by the NTSB (National Transportation Safety Board) re-veals several factors that might interfere with pilots' abilities to act properly when faced with emergency landings. The first of these is a reluctance to accept the emergency situation which may ini-tially paralyze the pilots, and their reaction may be a failure to lower the nose to achieve the best glide speed, delay in selecting a suitable landing site, or indecision in general. Often the first step in this denial reaction is to think, "This can't be happening to me!" Remember, when panic sets in, the thinking process shuts down and logic takes a hike. Pilots may quit flying the airplane and be-come passengers as the airplane takes over and they permit it to fly them.

In an effort to minimize damage to the airplane, pilots may be overly concerned with attempting to save the aircraft and neglect the basic rules of airmanship. Often pilots' financial stakes in the aircraft or the sometimes mistaken belief that an undamaged airplane guarantees no bodily harm can handicap the pilots. Simply from a physics per-spective, it is better that potentially lethal energy be dissipated by the aircraft rather than its occupants, even if it means that some air-craft damage will result. Also, pilots' fear of injury can engender panic, which is likely to override all but instinctive reactions to the impending situation. It is under just such conditions that pilots might allow a plane to stall at too high an altitude above the ground in an effort to touch down at as slow a speed as possible.

When dealing with forced landings proper training, beforehand, will make an enormous difference in the successful outcome of the real thing. There is an old joke that for some pilots, every landing is an emergency. However, even a real emergency landing can be treated as a normal, routine event, provided that pilots are properly trained and prepared for an unplanned landing. (See "You're Lucky to be Alive! above for an account of a routine off-airport, unplanned landing.) Always expect the unexpected and there will be no surprises, and that's our objective—to avoid being surprised.

## CFIT

The surprising thing about CFIT (controlled flight into terrain) is that it is really controlled. Highly experienced pilots (and crews with more than one experienced pilot aboard) fly the airplane right into the ground (or worse, a mountainside), usually at night and usually during the course of executing a nonprecision approach to an airport located in hilly country. And they just seem to keep doing it. This sort of activity is almost invariably fatal to all aboard.

Although it has always been with us, CFIT has only fairly recently been defined. Formerly this sort of thing was just another unexplained accident, but now efforts are being made to explain this phenomenon and to correct its causes. A great deal of attention is now being placed on the phenomenon known as CFIT. And progress is being made in explaining its causes.

Although as mentioned above, it happens to highly experienced pilots and crews, you don't have to be an experienced pilot to fly an airplane right into a mountainside or into the ground. Anybody can do it. It doesn't have to be at night, although it usually is, and it doesn't have to be on an approach. It also happens shortly after takeoff in cloud, day or night. It doesn't even have to be in IMC, although it usually is. Are you getting the message? It can happen to anybody in any conditions of flight. All it takes is a little lapse on the part of pilot or crew.

Brian Jacobson, in his book *Flying on the Gages*, relates the following typical example of CFIT:

*One major reason for following your position as the flight progresses is to keep yourself clear of rising terrain. Controlled flight into terrain has always been a problem for instrument aircraft, and those piloted by professionals and nonprofessionals alike have been a victim of this phenomena. One major accident happened when the crew of a DC-8 decided to hold over Salt Lake City while trying to solve an electrical fault that had taken much of the airplane's electrical equipment off line. The crew allowed the airplane to drift into an area of high terrain, outside its assigned holding pattern, where it crashed. Controllers were trying to reach the crew to warn them, but the pilots were off the frequency talking to their maintenance base on their only available radio when the accident took place.*

*This accident should serve to remind you about situational awareness and proper navigation. The DC-8 crew's awareness was not satisfactory. In effect, no one was flying the airplane because none of the crew members were aware of its position. And when contact with controllers was broken to discuss the problem with company maintenance personnel, the crew was totally responsible for knowing where the airplane was and that it was clear of the high terrain, but none of the pilots did the job. Perhaps they were overly reliant on radar to keep them out of trouble, or they just got so involved in their problem that none of the pilots realized that no one was flying the airplane.*

It boils down to a case of situational awareness. Not only must pilots be aware of their spatial relationship with Mother Earth (in terms of keeping the airplane upright), but they must also know precisely where they are in terms of the surrounding geography at all times. CFIT can only occur when the pilots, for whatever reason, lose this awareness. It may be the result of a distraction as in the above case (Remember the airliner that flew right into the Everglades while the entire crew was distracted?), or it might be simple brain fade, but whatever it is, it is extremely costly.

One factor is always present whenever there is a CFIT accident. The airplane was outside of protected airspace before it hit, and, of course, when it hit. Thus, if you remain in protected airspace it will never happen to you.

## The 609 ride

When a pilot certificate or rating is issued, the recipient is told, or should be told, that the administrator not only has the authority to issue such certificates and ratings, but if the administrator has reason to question the competence of the holder of such certificate or rating, the holder may be asked to demonstrate his or her competence by taking a checkride with an FAA inspector to determine if in fact the holder is indeed competent. The authority for this action is the Federal Aviation Act of 1958, Section 609, which is codified as 49 U.S.C. (United States Code), section 44709, hence the term "609 ride."

The NTSB (National Transportation Safety Board) and its ALJs (administrative law judges) have consistently held that the dear old FAA must have a valid reason for requiring an airman to undertake such an ordeal. In other words, an airman can't be arbitrarily called in to take a 609 ride. What kind of reason must the FAA have for calling into question the competence of an airman? Well, if you become involved in an accident, your competence is surely questionable, or it wouldn't have happened in the first place. If you are involved in an incident, it may be because you are incompetent to exercise the privileges of your certificate and ratings thereon, and your competence is surely questionable if you commit a violation, isn't it?

Thus, a 609 ride almost always follows an accident, or a violation, and frequently follows an incident. Writing in the *Journal of the Lawyer-Pilots Bar Association* (Spring 1991), Alan Armstrong says:

> *If a pilot lands gear up, lands on the wrong runway, or if his aircraft sustains damage due to alleged improper procedures or techniques, his competence may be challenged by the FAA. Following such an incident or accident, the investigating inspector may, at a minimum, require the airman to submit to a 609 re-exam. Depending on the circumstances surrounding the incident or accident, and/or the extent to which either the pilot or his representative is able to mollify any concern or anger held by the inspector, a successful 609 ride ... may conclude the FAA's interest in the matter.*
>
> *Air carrier and air taxi pilots, together with flight instructors, may find themselves taking 609 rides even though they have not been involved in operational errors. A professional pilot*

*who busts several periodic flight checks required by Parts 121
or 135 may find the Agency questioning his competency to be
an airman or to hold an instructor rating.*

*Further, an instructor's student who commits a serious act of
aeronautical ineptitude may find the FAA is interested both
in his teaching techniques and/or piloting capabilities.*

The ride itself is usually not particularly difficult for a competent pilot. It is designed to re-examine the airman on the specific task or
tasks of airmanship which have been called into question. For example, if there is a mishap during the course of a crosswind landing,
and an airplane gets bent or it otherwise comes to the attention of
the FAA, and the pilot should be required to submit to a 609 reeval-
uation, the pilot's ride would likely consist of a demonstration of one
or more crosswind landings.

I am familiar with a case in which a primary flight instructor had
such an atrocious bust rate (the poor fellow couldn't even get as
many as half his private applicants through their checkrides satisfactorily) that the FAA ordered him to come in for a re-evaluation of his
ability as an instructor. It took him four trips to the local FSDO to regain his privileges. The first two times he didn't even get to fly. His
knowledge of instruction techniques was so bad as demonstrated on
oral quizzing that he was sent home for more study. Then, when he
did finally get to fly, it took him two tries to satisfy the inspector.
One wonders how he acquired his instructor certificate in the first
place. (He had gone through a "diploma mill" and received his certificate from their tame staff examiner. That school was ultimately
shut down.)

Again in 1996 Alan Armstrong wrote on the same subject for the
quarterly publication of the LPBA (*The Lawyer-Pilots Bar Association
Journal*, Summer 1996). In an article entitled, "What Is A Reasonable
Basis For Requesting A Section 609 Re-Examination?" after the re-
codification of the act (changing 49 U.S.C. App. sec. 1429 to 49
U.S.C. 49 section 44709), he had this to say, "(The) statute authorizes
the Administrator to 're-examine an airman holding a certificate...,'
but the statute does not give any guidance in terms of under what
circumstances a request for reauthorization is appropriate." He goes
on to cite several cases decided by the National Transportation
Safety Board (the "Board"), which seem to throw some light on what
is deemed to be "reasonable." And although the FAA has repeatedly

attempted to exercise its authority in what would appear to the average citizen as totally unreasonable, the Board has fairly consistently ruled in favor of the airman in determining just what is reasonable.

Thus, if you are ever asked to visit your local FSDO for the purpose of taking a 609 ride and you believe there is no reasonable basis for calling your competence into question, you can refuse to do so and you have a pretty good chance of winning when you appeal to the NTSB. And if you do win on the basis of the FAA having been unreasonable, your costs might be recoverable under the EAJ (Equal Access to Justice) act.

# Glossary

**3LMB** Three light marker beacon receiver

**A&P** Mechanic certificate for airframe and power plant (formerly A&E for airframe and engine but with the advent of jets, it became A&P) Also refers to the individual holding such a certificate

**AC** Advisory circular

**A/C** Aircraft

**ACDO** Air Carrier District Office

**AD** Airworthiness directive (Issued by the FAA requiring compliance to ensure the continued safety of a specific make and model of aircraft)

**ADF** Automatic direction finder (receiver in the aircraft)

**AFM** Approved flight manual—required to be on board an aircraft for it to be deemed airworthy

**AGI** Advanced ground instructor

**AGL** Above ground level

**AI** Authorized inspector (Mechanic with inspection authorization)

**AIM** Airman's information manual

**Airman** Person who holds a FAA certificate

**ALJ** Administrative law judge

**ALPA** Airline Pilots Association

**AME** Aviation medical examiner—Doctor who has the authority to issue medical certificates to persons seeking or holding FAA certificates

**AOPA** Aircraft Owners and Pilots Association

**APPM** Accident prevention program manager (see SPM)

**APS** Accident prevention specialist (see SPM)

**ARTCC** Air route traffic control center

**ASI** Airspeed indicator. Also aviation safety inspector (see inspector)

**ASOS** Automatic surface observing system

**ATC** Air traffic control—all of it

**ATCT** Air traffic control tower

**Basic empty weight** Weight of empty airplane without fuel.

**Behind the power curve** (or area of reverse command). Slowest speed at which an airplane can sustain level flight with full power. In order to gain altitude (climb), the pitch attitude must be reduced (the nose of the airplane lowered)

**BFR** Biennial flight review

**BGI** Basic ground instructor

**CA** Commercial-airplane (pilot certificate)

**CAMI** Civil Air Medical Institute

**CAP** Civil Air Patrol

**CAVU** Ceiling and visibility unlimited (really great weather)

**Certificate** Issued by the FAA, conferring airman privileges. It will have one or more ratings placed on it delineating specific privileges

**CFI-ASE** Certified flight instructor—Airplane single engine

**CFII** Certified flight instructor—Instrument

**CFIME** Certified flight instructor—Multiengine

**CFIT** Controlled flight into terrain

**CFR** Code of federal regulations (Title 14 of the CFR is the body of law that governs all things aeronautical, and is frequently erroneously referred to as FARs)

**CG** Center of gravity (point on which an airplane in flight is balanced)

**CLC** Course line computer (original term for RNAV)

**CNS** Central nervous system

**CPU** Central processing unit

**CRM** Cockpit resource management

**CRT** Cathode-ray tube

**DG** Directional gyro (formerly called the gyro compass and now known as the heading indicator)

**DME** Distance measuring equipment

**DPE** Designated pilot examiner

**EAA** Experimental aircraft association

**EAJ** Equal Access to Justice Act—Entitles a person to recover his/her legal costs incurred in defending an unfounded violation charge by the FAA

**EFAS** Enroute Flight Advisory Service also known as Flight Watch

**EFC** Expect further clearance (time)

**EGT** Exhaust gas temperature gage

**FAA** Federal Aviation Administration (the agency)

**FAF** Final approach fix

**FAR** See federal aviation regulations

**FBO** Fixed base operator

**Federal aviation regulations** see CFR (FAR is the colloquial term used instead of the proper term, CFR)

**Flight Standards** Division of the Agency governing the activities of Airmen

**Flight Watch** (see EFAS)

**FSS** Flight service station

**FSDO** Flight Standards District Office

**GA** General aviation—also go-around

**GADO**  General Aviation District Office

**GAHA**  General Aviation Historical Society

**GAMA**  General Aviation Manufacturer's Association

**GPS**  Global positioning system (satellite based)

**GS**  Glideslope

**Gyro compass**  see DG

**Heading indicator**  see DG

**High-altitude structure**  Airspace above 18,000 feet and below the troposphere

**HUD**  Heads-up display

**IA**  Inspection authorization

**IAC**  International Aerobatic Club

**IAF**  Initial approach fix

**IAP**  Instrument approach procedure

**ICAO**  International Civil Aeronautics Organization

**ICC**  Instrument competency check

**IFR**  Instrument flight rules

**IGI**  Instrument ground instructor

**ILS** Instrument landing system

**IMC** Instrument meteorological conditions (Flight by reference to the flight instruments—in cloud)

**Inspector** Aviation safety inspector, operations, general aviation and air carrier (flight operations), airworthiness, general aviation and air carrier (maintenance), avionics, and safety program manager (formerly accident prevention specialist, and then accident prevention program manager). (The FAA is big on titles)

**IRA** Instrument rating—Airplane

**LOC** Localizer

**LOI** Letter of investigation—sent to airperson suspected of having committed a violation of one or more regulations. It is the invitation for the airperson to hang him/herself by making an admission against his/her interest. Although it is implied, the airperson is under no compulsion to answer.

**LOM** Outer compass locator (nondirectional beacon colocated with an outer marker).

**Loran C** Long-range navigation

**MAC** Mean aerodynamic chord (imaginary straight line running from the leading edge to the training edge of a wing). Not to be confused with a great big hamburger as one applicant answered when asked to define MAC.

**Max gross** Maximum allowable weight of fully loaded airplane

**MDA** Minimum descent altitude (As low as you can go on a nonprecision approach without being in position to make a normal landing and having the runway of intended landing in sight)

**MEA** Minimum en-route altitude

**MEL** Multiengine land, also minimum equipment list

**MM** Middle marker

**MSL** Mean sea level

**MVA** Minimum vectoring altitude

**NDB** Nondirectional beacon (ground transmission station that drives the ADF receiver in the aircraft)

**Ninety-nines** International organization of women pilots

**NORAD** No radio (an aircraft without radio communication)

**NOTAM** Notice to airman

**NPRM** Notice of proposed rulemaking

**NTSB** National Transportation Safety Board

**OAT** Outside air temperature gage

**OBS** Omni bearing selector

**OJT** On-the-job training

**OM** Outer marker

**PA** Private-airplane (pilot certificate)

**PIC** Pilot in command—both "acting" (the person driving the airplane) and "logging," any other(s) authorized to log PIC time on a specific flight

**PIREP** Pilot report (Weather conditions forwarded by pilots for the benefit of other pilots)

**POH** Pilot's operating handbook (not to be confused with the approved flight manual). A required document that must be on board the airplane

**PTS** Practical test standards

**RAPCON** Radar approach control

**Rating** Privileges specified on a pilot certificate

**REILS** Runway-end identifier lights

**RNAV** Direct routing by means of the course line computer, Loran C, or GPS

**RTO** Rejected takeoff

**SAR** Search and rescue (A CAP term)

**SB** Service bulletin (Issued by the manufacturer suggesting certain safety-related changes in its product)

**SIC** Second in command (copilot or more properly, nonflying pilot)

**SID** Standard instrument departure

**SODA** Statement of demonstrated ability, commonly referred to as a "waiver." Issued to an airman (airperson) who fails to meet the medical standards for his/her grade of certificate but demonstrates the ability to operate at that level despite the deficiency. (Color-blindness, amputated arm, missing one eye, etc.)

**SPM** Safety program manager (an ASI in each FSDO assigned the duty of preparing and presenting safety programs)

**SSA** Soaring Society of America

**Stall** An aerodynamic stall occurs when the angle of attack exceeds a critical point

**STAR** Standard terminal arrival route

**STC** Supplemental type certificate (authority to modify an existing, approved manufactured product)

**TRACON** Terminal radar approach control

**Useful load** Everything between basic empty and max gross weights

**VFR** Visual flight rules

**VMC** Visual meteorological conditions (flight by outside references)

**Vmc (or Vmca)** Speed below which a twin-engine airplane is unable to maintain directional control with the critical engine inoperative and the working engine developing maximum power, along with several other adverse factors (definitely a bad situation in which to find oneself)

**Vne** Never exceed speed (red radial on the ASI, also known as redline)

**Vno** Normal operating airspeed range—Green arc on the Airspeed Indicator

**VOR** Very high frequency omnirange

**W&B** Weight and balance

**Zero-zero** Ceiling zero, visibility zero (really grim weather, usually fog)

# Bibliography

AC 61-21A (Flight Training Handbook). U.S. Department of Transportation. Federal Aviation Administration. (Available from the GPO, or reprinted by ASA Publications)

Buck, Robert N. *FLYING KNOW-HOW.* Delacorte, 1975.

Buck, Robert N. *Weather Flying.* New York: MacMillan, 1970.

Fried, Howard J. *Flight Test Tips & Tales from the Eye of the Examiner.* New York: McGraw-Hill, 1996.

Jacobson, Brian. *Flying on the Gages.* Union Lake, MI: Odyssey Aviation Publications, 1996.

Kershner, William K. *The Advanced Pilot's Flight Manual* (current edition). Ames, Iowa: Iowa State University Press, 1970.

Langewiesche, Wolfgang. *Stick and Rudder.* New York: McGraw-Hill, 1995.

Lowery, John. *Anatomy of a Spin.* Airguide Publications, Inc.

Ploudre, Harvey S. *The Compleat Taildragger Pilot.* Published by the author.

# Index

# About the Author

Howard Fried (of Michigan) is a veteran pilot with more than 40,000 hours of flight time. He is the author of the book *Flight Test Tips and Tales*, and is a regular columnist ("Eye of the Examiner") for *Flying*, the world's most widely read aviation magazine. He is a flight instructor and former FAA pilot examiner.